X-RAY AND
ELECTRON METHODS
OF ANALYSIS

PROGRESS IN
ANALYTICAL CHEMISTRY
Based upon the Eastern Analytical Symposia

Series Editors:

Ivor L. Simmons
M&T Chemicals, Inc., Rahway, N. J.

and Paul Lublin
General Telephone and Electronics Laboratories, New York, N.Y.

Volume 1
H. van Olphen and W. Parrish
X-RAY AND ELECTRON METHODS OF ANALYSIS
Selected papers from the 1966 Eastern Analytical Symposium

In preparation:
Volume 2
E. M. Murt and W. G. Guldner
PHYSICAL MEASUREMENT AND ANALYSIS OF THIN FILMS
Selected papers from the 1967 Eastern Analytical Symposium

Volume 3
K. M. Earle and A. J. Tousimis
X-RAY AND ELECTRON PROBE ANALYSIS
                    IN BIOMEDICAL RESEARCH
Selected papers from the 1967 Eastern Analytical Symposium

PROGRESS IN ANALYTICAL CHEMISTRY
VOLUME 1

X-RAY AND ELECTRON METHODS OF ANALYSIS

Edited by **H. van Olphen**
National Research Council
Washington, D. C.

and **William Parrish**
Philips Laboratories
Briarcliff Manor, N. Y.

Ψ Springer Science+Business Media, LLC 1968

Library of Congress Catalog Card Number 68-13392

ISBN 978-1-4899-5915-7   ISBN 978-1-4899-5913-3 (eBook)
DOI 10.1007/978-1-4899-5913-3

© *1968 Springer Science+Business Media New York*
Originally published by Plenum Press in 1968.
Softcover reprint of the hardcover 1st edition 1968

PREFACE

The policy of the Eastern Analytical Symposium to have only invited speakers leads to the expectation of first-rate papers, and the idea of publication becomes very appealing. For many years, people have asked whether there would be publication of the proceedings, but the wide range of topics covered by the Symposium makes for an unwieldy volume. However, this year it was decided to start a series which will be limited each year to selected sessions. Chosen from the 1966 Symposium Program were "X-ray and Electron Methods for Chemical Analysis," chaired by Dr. W. Parrish, and "Scattering Techniques for the Determination of Surface Area, Porosity, and Particle Size," chaired by Dr. H. van Olphen. These sessions are related by the electron, optical, and X-ray methods of chemical and structural analysis employed.

The task of editing the papers presented at the two sessions involved was undertaken by the respective session chairman, without whose drive and cooperation the task of producing this first volume would have been insurmountable. It is hoped that this publication will be the precursor of many volumes to come and that there will be no undue delay from the date of presentation to the date of publication.

Ivor L. Simmons
Paul Lublin

CONTENTS

Contents

Chapter VII
The Analysis of Low-Angle Light Scattering from Simple Mixtures ..
by Daniel Caulfield, Yung-Fang Yao, and Robert Ullman

I. X-RAY DIFFRACTOMETRY METHODS FOR COMPLEX POWDER PATTERNS*

William Parrish

Philips Laboratories
Briarcliff Manor, New York

Complex X-ray powder diffraction patterns arise from low crystallographic symmetry and/or large unit cell dimensions, multiple phase specimens and structural imperfections. This paper deals with some recent advances in the techniques of recording complex powder patterns. The increased dispersion arising from the use of longer wavelength radiation (e.g., Cr $K\alpha$, $\lambda = 2.28$ Å) greatly reduces the difficulties caused by overlapping reflections. The experimental conditions are selected to achieve the optimum compromise between high intensity, peak-to-background ratio, and resolution. High angular precision is required for computer indexing and is obtained by step scanning around the peak and fitting a parabola to the uppermost points. A fine focus X-ray tube is employed and the anode viewed at relatively large angles ($>10°$) to reduce self-absorption; high specific target loading and thin Be windows provide high intensities. A simple vacuum chamber eliminates most of the air-path absorption losses. The specimen has a fixed curvature corresponding approximately to that appropriate for the lowest 2θ-angle to be scanned, thereby reducing the flat specimen aberration. The X-ray optics in the axial plane are lengthened to further increase the intensity without loss of resolution. The techniques are illustrated by examples of analyses of steroids and low symmetry minerals.

1. INTRODUCTION

X-ray powder diffractometry is an indispensable method for phase identification and structural characterization of crystalline materials. In recent years it has become necessary to analyze materials whose patterns are complex and difficult to interpret because they contain a large number of closely spaced and overlapping lines. The difficulties increase with the

---

* This research was supported in part by the Advanced Research Projects Agency and was technically monitored by the Air Force Office of Scientific Research under Contract No. AF49(638)–1234.

number of crystallographically distinct phases in the specimen, decreasing crystallographic symmetry, and increasing unit cell dimensions. This is illustrated in figure 1 by the three principal minerals in a granite. The superposition of the rhombohedral quartz, monoclinic orthoclase, and triclinic albite patterns produces a complex composite pattern. The quantitative X-ray petrological analysis of such samples is difficult using Cu $K\alpha$ radiation. It will be shown that longer wavelength Cr $K\alpha$ radiation increases the dispersion and is of considerable help in reducing the number of overlaps.

The desired analytical information often extends beyond identification and may require determination of the degree of substitutional isomorphism, the distribution of different atoms on equivalent lattice sites, accurate lattice parameters, automatic quantitative analyses of large numbers of samples, and similar analyses with a precision that may be near the limit of present instrumentation and methods. The success of such analyses depends in large measure on proper planning, which may require a considerable amount of preliminary theoretical and experimental study to achieve optimum conditions.

This paper describes some recent advances in reflection specimen powder diffractometry, including the use of a single curvature specimen to reduce the flat specimen aberration, the advantages of long parallel slit assemblies to limit axial divergence, techniques of obtaining high intensity Cr $K\alpha$ radiation, optimum conditions for operating the X-ray tube and the dependence of intensity and background on voltage and angle of view of the anode. Some steroid patterns are analyzed, and the results are compared with powder camera data. It is assumed that the specimens are well crystallized so as to avoid the additional complications of line broadening and shifts arising from imperfect structures, disorder, small crystallite size, etc. It is evident that the analysis of profiles distorted by the crystal structure of the specimen requires a knowledge of the instrumental broadening factors to separate these different contributions.

Transmission specimen diffractometer methods require a focusing crystal monochromator and are not so well developed. The transmission method has many advantages and is extremely useful for measuring large d-spacings and as a supplementary technique to the reflection method. The development of focusing crystal monochromators which eliminate the $K\alpha_2$ line with only a small loss of $K\alpha_1$ intensity would further greatly enhance the capabilities of both methods.

2. DIFFRACTOMETER GEOMETRY

The ray diagram in the focusing plane of the conventional diffractometer[1,2] is shown in figure 2. The focal line of the X-ray tube F is the

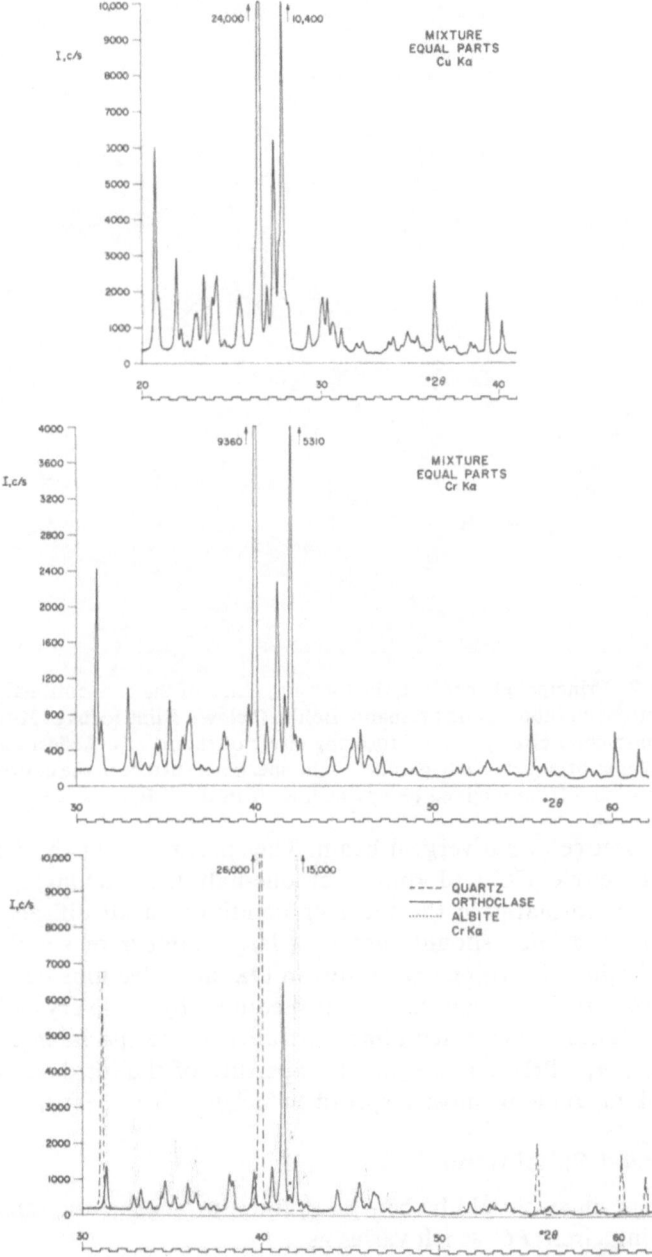

Figure 1. Complex powder pattern of the three principal minerals in a granite. Increased dispersion obtained by using longer wavelength X-rays aids in decreasing the number of overlapping reflections.

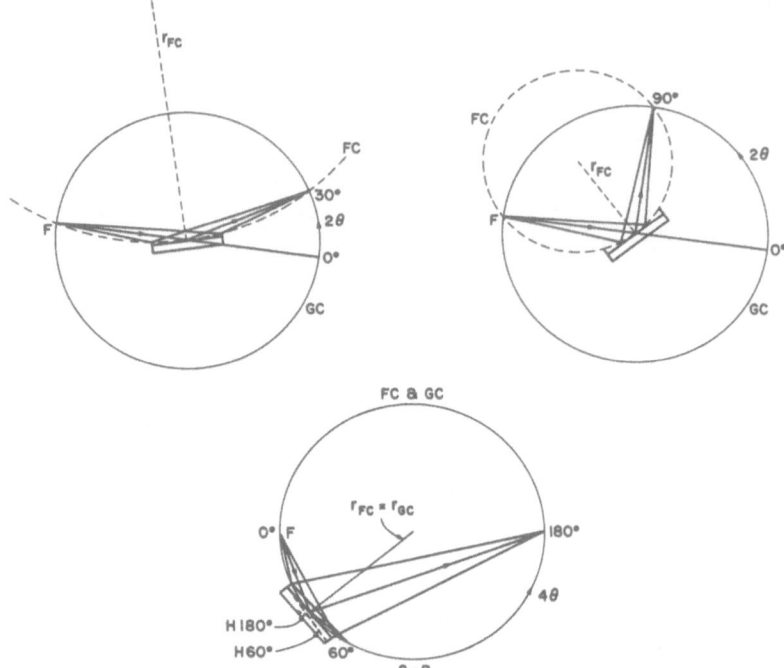

Figure 2. Principal elements in the focusing plane of the conventional powder diffractometer (above) and Seemann–Bohlin (below). *F* line focus of X-ray tube, *GC* goniometer circle, and *FC* focusing circle of radius r_{FC}. Reflection occurs from planes oriented nearly parallel to the specimen surface in the conventional geometry and from inclined planes such as *H* in the S-B.

geometric source of a divergent beam. The specimen is at the center of the goniometer circle *GC* and rotates at one-half the angular speed of the receiving slit to maintain the focusing condition at all diffraction angles. Ideally, the specimen should contain a large number of small, randomly oriented, highly absorbing crystallites so that all reflections occur from the top surface. The Bragg law can be satisfied only by those crystallites whose reflecting planes are oriented almost parallel to the specimen surface, i.e., within $\frac{1}{2}\alpha + \tau$, where α is the angular aperture of the incident beam and τ is the rocking angle or mosaic spread (whichever is larger).

2.1. Curved Specimens

The specimen should be bent so that its surface has the same radius as the focusing circle *FC* which varies as

$$r_{FC} = r_{GC}/2 \sin \theta \tag{1}$$

and hence the curvature must change continually during the scan.

Flat specimens have been most widely used in diffractometry and cause an aberration in the focusing. The line profiles are broadened asymmetrically and shifted to smaller diffraction angles, thereby introducing a systematic error in the positions of the centroids:

$$\Delta 2\theta_{FS} = -(l^2/12r^2_{GC}) \sin 2\theta$$
$$= -(\alpha^2/6) \cot \theta \tag{2}$$

where $\Delta 2\theta_{FS}$ and α are expressed in radians. This expression is shown as a dashed curve in figure 3 for $\alpha = 1°$ and $r_{GC} = 185$ mm. The shift is zero at $2\theta = 180°$ and increases at small diffraction angles where r_{FC} approximates a flat specimen. This apparent contradiction can be understood from the fact that when using a fixed α the length of specimen irradiated l also increases with decreasing θ:

$$l = \alpha r_{GC}/\sin \theta \tag{3}$$

where α is in radians. The deviation h from the focusing circle of incident rays striking the specimen at a distance $l/2$ from the center is

$$h = r_{FC} - [r^2_{FC} - (l^2/2)]^{1/2} \tag{4}$$

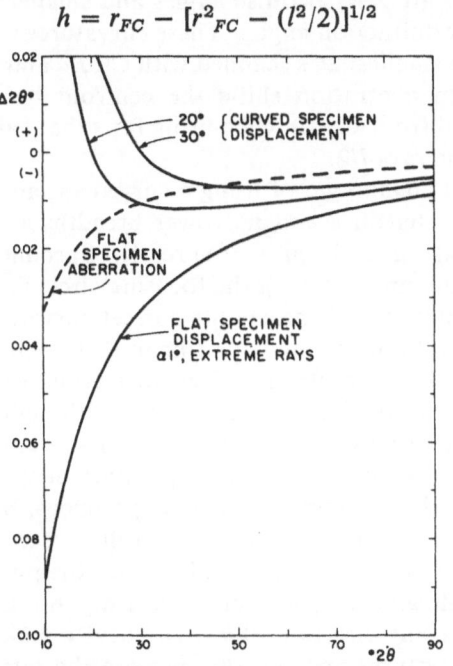

Figure 3. Displacement of centroid caused by flat specimen aberration (dashed line), and the displacements calculated for the extreme rays of flat and curved specimens for $\alpha = 1°$, $r_{GC} = 185$ mm.

If a flat or curved specimen is displaced radially a distance s from the focusing circle the entire line profile is shifted:

$$\Delta 2\theta_{SD} = (2s/r_{GC}) \cos \theta \qquad (5)$$

where $\Delta 2\theta_{SD}$, expressed in radians, may be positive if the displacement is inside FC and negative if the displacement is outside FC. If s is small and the instrument is properly aligned there is no change in the shape of the profile. This shift is the most common source of systematic errors in d's.

The flat specimen aberration may be interpreted as a continuous series of specimen surface displacements in which h is equivalent to s. In the case of a specimen with no displacement, $h = s = 0$ at the center of the specimen and the central ray reaches the receiving slit at the correct 2θ angle. The extreme divergent incident rays strike a flat specimen below the focusing circle and reach the receiving slit at smaller 2θ angles. Figure 3 shows $\Delta 2\theta_{SD}$ for a flat specimen in which the values of $l/2$ used in eq. (4) vary with θ as shown in eq. (3). If the same calculation is made for specimens whose curvatures match the focusing circle at 20° and 30° (2θ) the displacements are zero at those angles and smaller than those of a flat specimen at other diffraction angles. These curvatures were selected because they are often the smallest 2θ's scanned with Cu $K\alpha$ and Cr $K\alpha$, respectively. The flat specimen aberration shifts the centroid by only one third the amount calculated from eq. (5) when using for s that value of h corresponding to the extreme rays $l/2$.

The practical advantages of using a curved specimen are illustrated by the higher peak intensities and narrower breadths at one-half ($W_{1/2}$) and one-tenth ($W_{1/10}$) peak heights (figure 4). Specimens with the same curvature, $r = 338$ mm matching the focusing circle at 31.8°, were used for both sets of profiles. This is nearly the correct curvature for quartz (10.0), but r should have been 155 mm for silicon (220). The (220) profile using $\alpha = 1°$ is practically free of flat specimen aberration because l is only 5 mm. When α is increased to 4°, $l = 21.8$ mm, and although there is some aberration present the resolution is much better for this curvature than for that obtained with the flat specimen. The preparation of specimens with a single curvature is virtually the same as for flat specimens, but the machining of the specimen holders is more difficult and will be described elsewhere.

Ogilvie[3] has described a device in which the powder is cemented on a flexible substrate and the curvature is varied continually during the scan to match the focusing circle at all θ's. The device makes it possible to use larger angular apertures and thereby increase the intensity without introducing flat specimen aberrations. However, the powder layer must be thin, and hence the reflected intensity for the same α is usually lower than that of the normal thick specimen preparations. If the incident beam penetrates

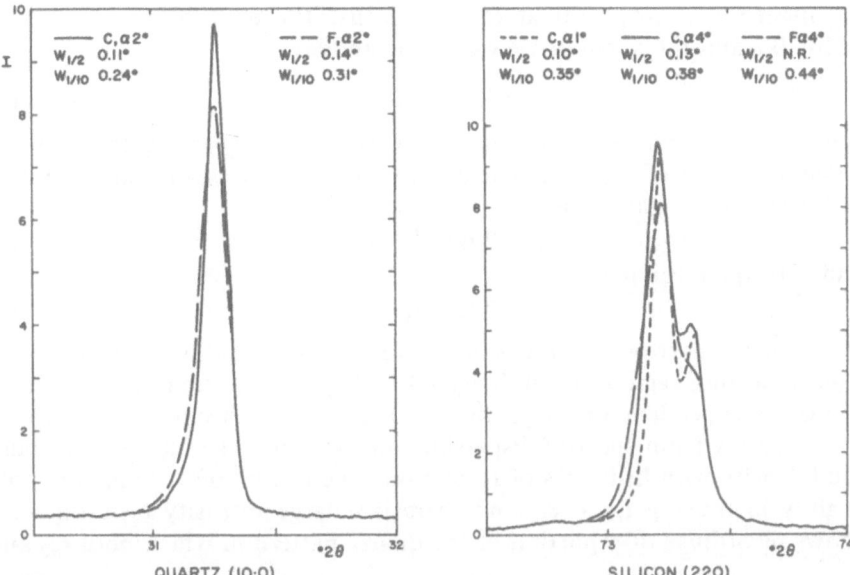

Figure 4. Effect of specimen curvature on line profiles. C curved specimen $r = 338$ mm (correct curvature at 31.8°), F flat specimen. Recorder scale increased by factor of four for silicon α 1° profile. Cr $K\alpha$, 40 kV, V filter, vacuum diffractometer, receiving slit 0.075 mm, 1/8°/min, time constant 4 sec. $W_{1/2}$ and $W_{1/10}$ are widths at one-half and one-tenth peak heights.

the powder layer, fluorescence and scattering from the substrate may increase the background, thus further reducing the peak-to-background ratio. This difficulty may be reduced by using a beryllium foil substrate in which there is a high degree of preferred orientation or a very thin elastic single crystal in which the reflecting planes are inclined to the surface.

It should be noted that curved specimens preclude the possibility of rotating the specimen in its own plane, a technique used to reduce the effect of crystallite size statistics on the relative intensities and preferred orientation on the specimen surface[1]. It thus becomes essential to use crystallites in the 5 to 10 μ region to minimize intensity errors arising from poor statistics in the number of reflecting crystallites when using curved specimens.

2.2. Intensity and Resolution

The projected widths of the source W_F and the receiving slit W_{RS} each adds a symmetrical broadening independent of diffraction angle. The widths must be narrow to achieve good resolution, but unfortunately the intensity is proportional to the widths. W_F depends on the actual width W_F' and the angle of view ψ of the anode:

$$W_F = W_F' \sin \psi \tag{6}$$

where ψ is defined with respect to the central ray passing through the goniometer axis of rotation. The aperture formed by a point at the center of the specimen and W_F is

$$\epsilon_F = 2 \tan^{-1}(W_F/2r_{GC}) \tag{7}$$

and with the receiving slit

$$\epsilon_{RS} = 2 \tan^{-1}(W_{RS}/2r_{GC}) \tag{8}$$

When a narrow line focus X-ray tube is used the line breadth in the front reflection region of a well-crystallized specimen free of strains etc. is primarily dependent on W_{RS} (figure 5). In back reflections the breadth arises mainly from spectral dispersion, and W_{RS} may be increased to raise the intensity with little loss of resolution. The resolution is improved only slightly in making $\epsilon_{RS} < \epsilon_F$, and there is a large intensity loss. Figure 6 shows recordings of a portion of the quartz pattern in which both ϵ_{RS} and

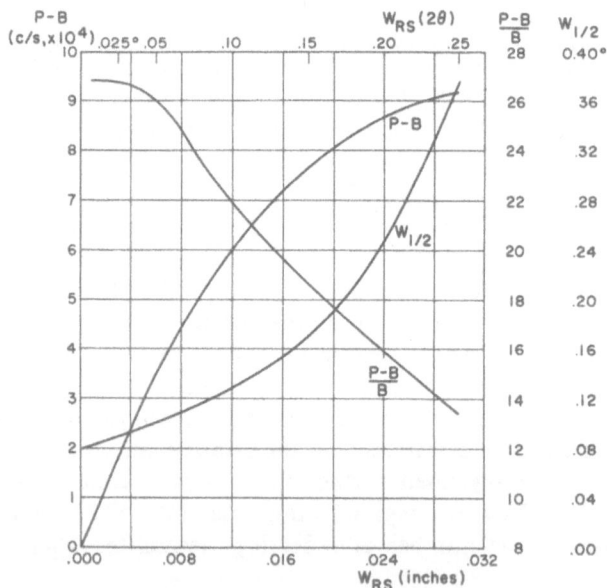

Figure 5. Effect of receiving slit width W_{RS} on peak intensity $P\text{-}B$, peak-to-background ratio $(P\text{-}B)/B$ and width at one-half peak height $W_{1/2}$. Curved silicon powder specimen $r = 338$ mm, (111), Cr $K\beta$, 38.8°, $\alpha = 2°$. Intensity measurements made with manual settings, profile recorded at $\frac{1}{8}°$/min and time constant 4 sec.

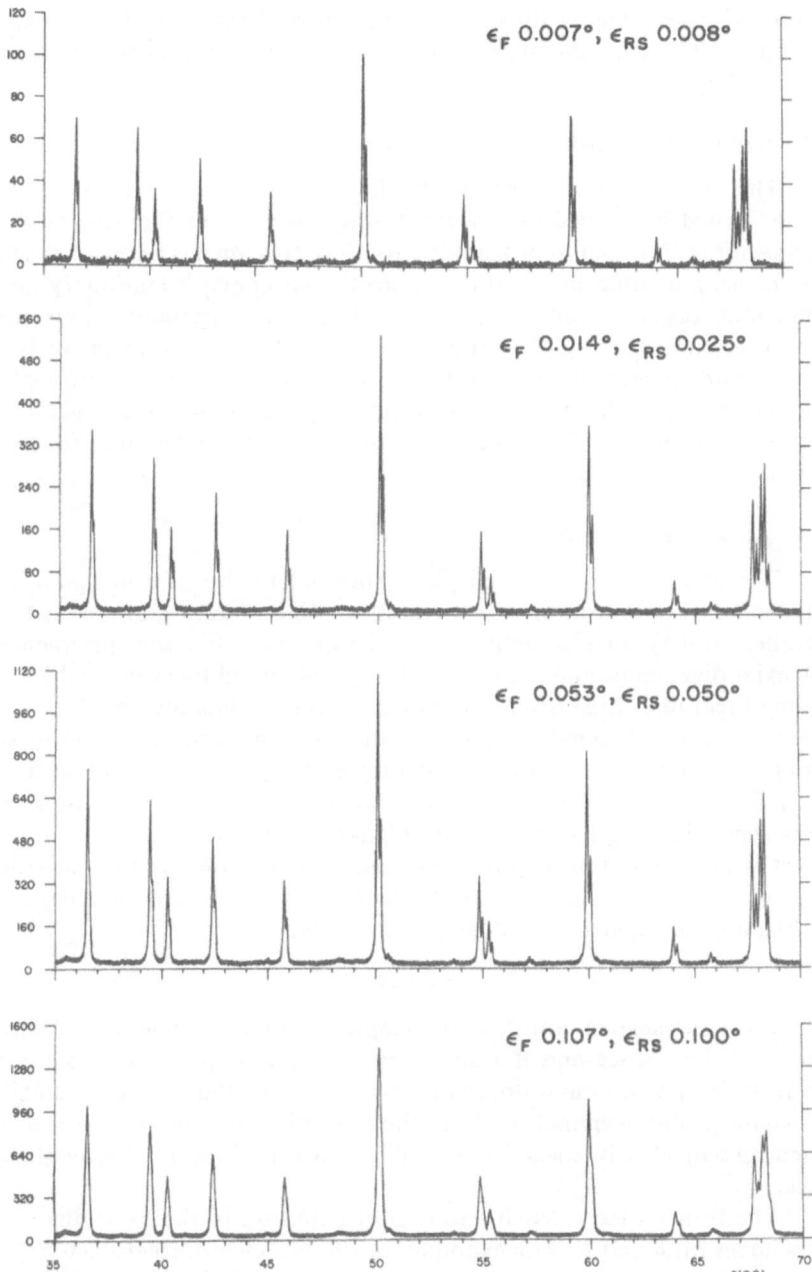

Figure 6. Decrease of resolution and increase of intensity as source width aperture ϵ_F and receiving slit aperture ϵ_{RS} are increased keeping $\epsilon_F \approx \epsilon_{RS}$. Quartz powder, Cu $K\alpha$ radiation, α 1°, $\frac{1}{4}$°/min, time constant 2 sec.

ϵ_F are increased, and it illustrates the reciprocal relation of intensity and resolution. The selection of apertures and their calculation has been recently described[4].

2.3. Seemann–Bohlin Diffractometer

The Seemann–Bohlin geometry shown at the bottom of figure 2 has not been widely applied in powder diffractometry. The focusing circle has a constant radius, and only a single curvature specimen is required to satisfy the focusing conditions at all θ's. Since the specimen is stationary and all reflections occur simultaneously, the design of specimen environment chambers for changing the temperature, pressure, or atmosphere is simplified, and several detectors may be used simultaneously. Although the angular range is limited, the method may be used advantageously in certain applications. The reader may consult recent publications for further details[5,6].

2.4. Axial Divergence

The apertures in the axial plane (normal to the focusing plane) also have a marked effect on the resolution, line shape, and intensity[1,2]. Parallel (Soller) slit assemblies placed before and after the specimen limit the axial divergence and are essential to obtain good line shape when using a long focal line. Line profile asymmetry results when divergent rays from the focus are diffracted by the specimen and detected at some distance along the axial direction from their origin. As the receiving slit scans over the reflection, the crossover radiation is detected on the low-θ tails in the front reflection region and on the high-θ tails in the back reflection. The effect is at a minimum in the mid-θ range. The profile is thus broadened asymmetrically and shifted from its correct position by amounts dependent on the angular aperture δ of the parallel slits:

$$\delta = 2 \tan^{-1} s/l_1 \tag{9}$$

where s is the spacing and l_1 is the length of the foils. The convolution of the axial divergence and flat specimen aberrations produces a systematic error in line position and dominates the shapes of the profiles at small θ's. When the profile asymmetry is large the resolution is reduced and the relative intensities of closely spaced lines will be in error due to the overlapping tails.

The transmitted intensity and its distribution in the axial direction is dependent on δ and l_1. For example, in the Norelco diffractometer[1] with $r_{GC} = 174$ mm, the distance between the focus and the far end of the first set of foils is 67.5 mm, and the axial length is 10 mm. If F is 8 mm long and

$\delta = 4.5°$, the seven central pairs of foils each see 4.8 mm of the focus and the remaining thirteen pairs see decreasing lengths, depending on their distance from the middle of the assembly. The distance from the ends of the foils near F to the far end of a 25-mm-long specimen is 132 mm. If the axial width of the specimen is 10 mm (parallel to the rotation axis), only the central pair of foils irradiates the entire width. Increasing the width to 15 mm permits radiation from the central eleven pairs to reach the entire focus. The distance between the end of the specimen near F to the end of the foils next to the antiscatter slit is 204.5 mm, and only the central pair can receive radiation from a specimen 16.1 mm wide. As a result of the overlapping divergent patterns, the intensity has a sharp peak along the median line[7] and some intensity is lost because of the limited dimensions of the X-ray optical system in this plane. The intensity may be increased 10 to 70% without loss of resolution by increasing the axial widths of the parallel slit assemblies, specimen, receiving, and antiscatter slits by amounts determined from a ray diagram.

Figure 7 shows the dependence of asymmetry on δ using good conditions for resolution (given in the legend). The peak intensities were normalized

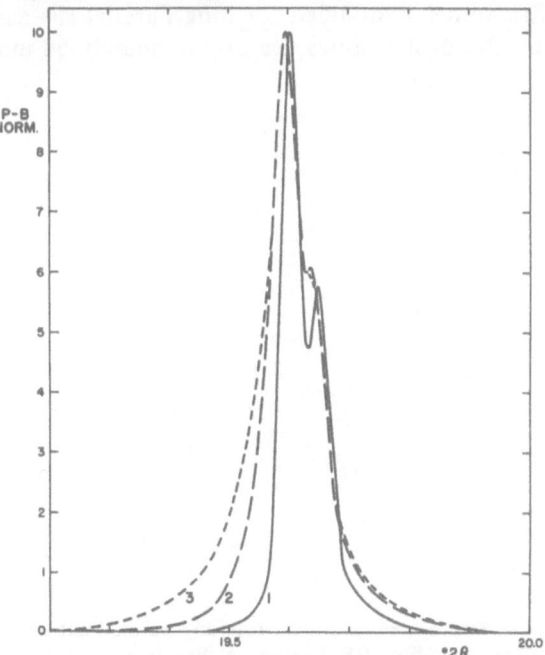

Figure 7. Effect of axial divergence on asymmetry of line profiles. Cu $K\alpha$, line focus 0.04 by 8 mm, W_{RS} 0.075 mm, curved $Pb(NO_3)_2$ specimen, (111) reflection, fixed time step scan 0.01° increments. Peak intensities normalized.

to facilitate comparison of the line shapes. The width of the profile at one-tenth peak height is a more sensitive measure than the width at one-half height because the lower portion of the tail is more affected by this aberration. Profiles 2 and 3 were obtained with the conventional 20-mm-long assemblies. Decreasing δ of both assemblies from 4.5° (profile 3) to 2.25° (profile 2) reduced $W_{1/10}$ from 0.27 to 0.21°, but the peak intensity was reduced 62%. Profile 1 was obtained using pairs of cylindrical slits[8] to simulate one section of a long assembly; for the incident beam, $l = 100$ mm, $\delta = 1°$; for the diffracted beam, $l = 148$ mm, $\delta = 1.6°$. $W_{1/10}$ was reduced to 0.13°, asymmetry was removed from the low-θ tail, and the resolution of the doublet was enhanced. The intensity was low, and it is not possible to compare it with other profiles because the experiment was equivalent to only one section of the assembly.

If the length and spacing of the foils are increased to maintain the same δ, the line symmetry is the same, the total intensity reaching the specimen is increased, and the distribution becomes more uniform. The smaller number of foils makes it possible to reduce δ without the intensity loss due to the sum of the foil thicknesses. Figure 8 is a schematic drawing of a powder diffractometer modified for long parallel slit assemblies. The foils indicated by the dashed lines are approximately 54 mm long in the

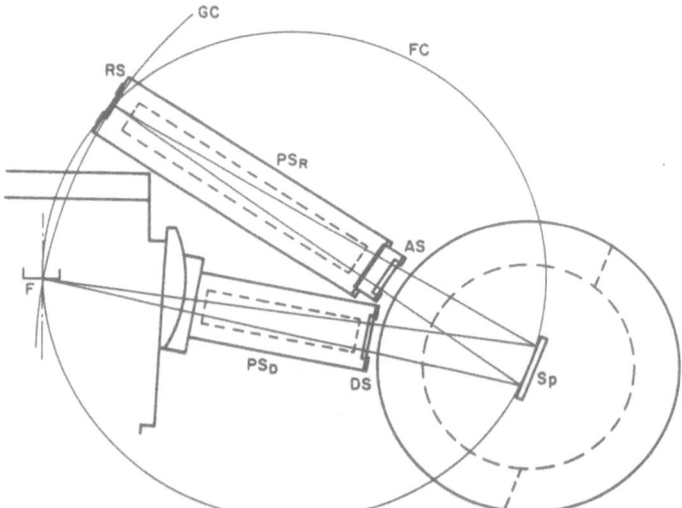

Figure 8. Diffractometer modified for long parallel slit assemblies. PS_D incident assembly, DS divergence slit, Sp specimen, AS antiscatter slit, RS receiving slit, PS_R receiving assembly. Foils indicated by dashed rectangles. Beryllium windows at ends of assembly and cylindrical chamber around specimen make it possible to pull vacuum.

incident beam assembly and 99 mm long in the diffracted beam. The divergence and antiscatter slits are close to the specimen, making it easier to eliminate unwanted scattered radiation. The maximum scanning angle is 157° for $r_{GC} = 185$ mm, $\alpha = 4°$, and $\psi = 10°$.

3. SELECTION OF X-RAY WAVELENGTHS

One of the most important steps in diffractometry technique is the selection of the most suitable wavelength and the proper operation of the X-ray tube to achieve maximum intensity and minimum background. A number of factors required to make these decisions are described in this section.

3.1. Dispersion

It has been shown that increasing the resolution by decreasing the slit apertures results in a loss of intensity and a consequent loss of precision or increased time for measurement. To minimize this difficulty the dispersion may be increased by using longer-wavelength X-rays.

The most widely used radiation for the powder method is Cu $K\alpha$, $\lambda = 1.54$ Å, which has adequate dispersion for many problems. Various difficulties are encountered in selecting suitable anodes for longer-wavelength X-rays. If we require that the important analytical region $d = 5$ to $1\cdot5$ Å falls within the maximum scanning angle of the diffractometer ($\approx 160°$), the longest wavelength that can be used is 2.95 Å. Ti $K\alpha$, $\lambda = 2.75$ Å, meets this requirement, but there is no $K\beta$-filter for this radiation, and hence a monochromator would be required. V $K\alpha$, $\lambda = 2.50$ Å, is a possibility with a Ti filter. The $L\alpha_1$ lines of Ag and Mo have too long wavelengths for this application. The longest-wavelength radiation readily available is Cr $K\alpha$, $\lambda = 2.29$ Å, but it has not been widely used because of low intensities. Recent improvements in X-ray tube design and the use of a vacuum path have made it possible to greatly increase the intensity.

The dispersion may be expressed as degrees θ per Angstrom:

$$-\Delta\theta/\Delta d = (114.59 \sin\theta \tan\theta)/\lambda \qquad (10)$$

and is plotted in figure 9 for Cu $K\alpha$ and Cr $K\alpha$ radiations. The d spacings 5 to 1.5 Å extend from 17.7 to 61.8° (2θ) using Cu $K\alpha$ and from 26.5 to 99.6° (2θ) using Cr $K\alpha$. It would require 66% more time to scan this range with Cr radiation while at the same scanning speed. This is compensated by the reflections occurring at larger 2θ angles, and hence a larger angular aperture may be used to achieve greater primary intensity. For $r_{GC} = 185$ mm, $\alpha = 1°$, $l = 22.3$ mm at 17.7°, and for the same l using Cr $K\alpha$,

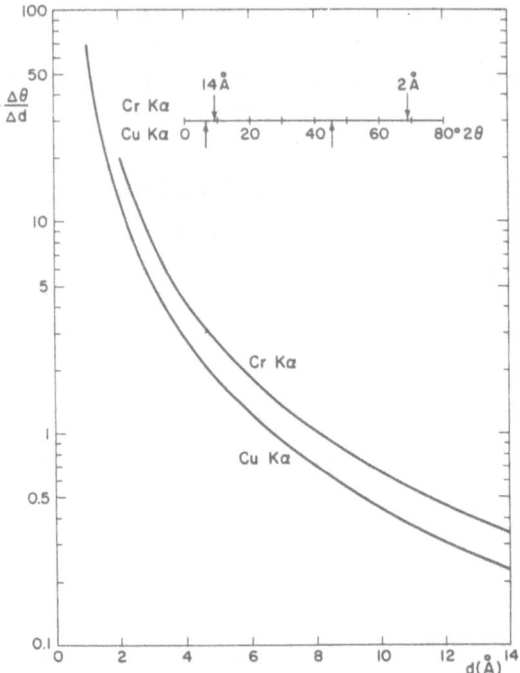

Figure 9. Dispersion curves for Cu $K\alpha$ and Cr $K\alpha$
radiations. $-\Delta\theta/\Delta d = (114.59 \sin \theta \tan \theta)/\lambda$.

α can be increased to 1.6°. The scanning speed thus can be increased by
nearly the same factor as the increase in scanning range and thereby obtain
the same counting statistical accuracy.

The Lorentz polarization factor for the powder diffractometer,

$$\text{L-p} = \frac{1 + \cos^2 2\theta}{\sin^2 \theta \cos \theta} \tag{11}$$

changes rapidly at small and large diffraction angles. At small diffraction
angles the observed intensities of the same d will be higher for Cu $K\alpha$ than
for Cr $K\alpha$, and at large diffraction angles the reverse is true. For $d = 5$ Å
the L-p is about 2.4 times greater for Cu $K\alpha$, and for $d = 1.2$ Å it is about
twice as high for Cr $K\alpha$.

Diffractometer patterns of the mineral topaz, $Al_2SiO_4(OH, F)_2$, space
group *Pnma*, $a = 4.65$, $b = 8.80$, $c = 8.39$ Å, provide an example of the
results of increased dispersion using the same resolution for Cu $K\alpha$ and
Cr $K\alpha$ (figure 10). The d range 4.4 to 1.6 Å using Cu $K\alpha$ extends from
20° to 57° and for Cr $K\alpha$, from 30° to 90°. The Cu pattern has about twice

the intensity, and there are no difficulties in indexing the pattern. The
d range 1.6 to 1.16 Å using Cu $K\alpha$ extends from 57° to 83°, and for Cr $K\alpha$,
from 90° to 158°. Both patterns have about the same intensity because a
larger aperture was used with Cr $K\alpha$. The effect of the larger dispersion can
be seen clearly, particularly in the regions with overlapping reflections.
Although the d range below 1.16 Å is not accessible with Cr $K\alpha$, there are
a sufficient number of high-angle reflections to provide precise lattice
parameters.

   Another advantage of Cr radiation is in the analysis of specimens
containing elements whose absorption edges occur at wavelengths greater
than that of the nickel β filter ($\lambda = 1.49$ Å) used with Cu radiation. The
incident continuous and characteristic line radiation with wavelengths less
than the absorption edges of the elements in the specimen cause those
elements to produce fluorescent X-rays which increase the background.
If the β filter is placed in the diffracted beam it will absorb fluorescence
radiation having wavelengths shorter than the absorption edge of the filter.
Cr radiation thus is ideal for ferrous materials because the K-absorption
edge of the V filter $\lambda = 2.07$ Å is just above that of Fe $K\alpha$ radiation
$\lambda = 1.94$ Å.

3.2. Spectral Distribution of Primary Beam

   The spectrum of an X-ray tube consists of characteristic lines super-
imposed on continuous radiation. The wavelengths of the lines are determined only by the anode element and the intensity by the applied voltage,
angle of view ψ of the anode, and transmission of the window. The continuous radiation changes its spectral distribution with voltage. Increasing
the voltage causes the spectrum to begin at shorter wavelengths and the
broad high-energy peak to increase in intensity. At the same voltage the
continuous radiation intensity increases with atomic number of the
anode.

   The ratio of the intensities of the characteristic line to the continuous
radiation can be measured with a scintillation counter which has nearly
uniform quantum-counting efficiency over the spectral range of interest[1].
It is essential to use small apertures (≈ 0.1 mm) such as those formed by
pairs of crossed rod slits[8] and separated by long distances (≈ 200 mm) to
reduce the intensity to the linear range of the circuits. The ratio can be
determined with a single channel pulse height analyzer with or without
balanced filters. Figure 11 shows pulse amplitude distributions from a
copper target X-ray tube; the intensities were recorded using fixed time, a
one dial unit (DU) window and 1 DU steps. In these plots the pulse
amplitude is proportional to the X-ray quantum energy and increases with
increasing DU so that the wavelength decreases to the right.

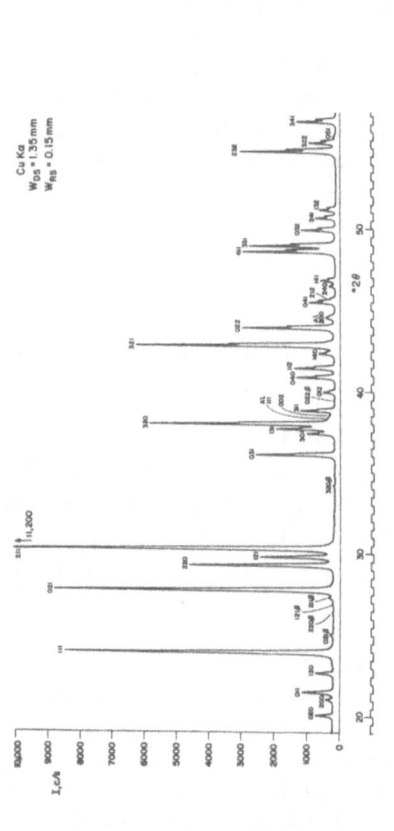

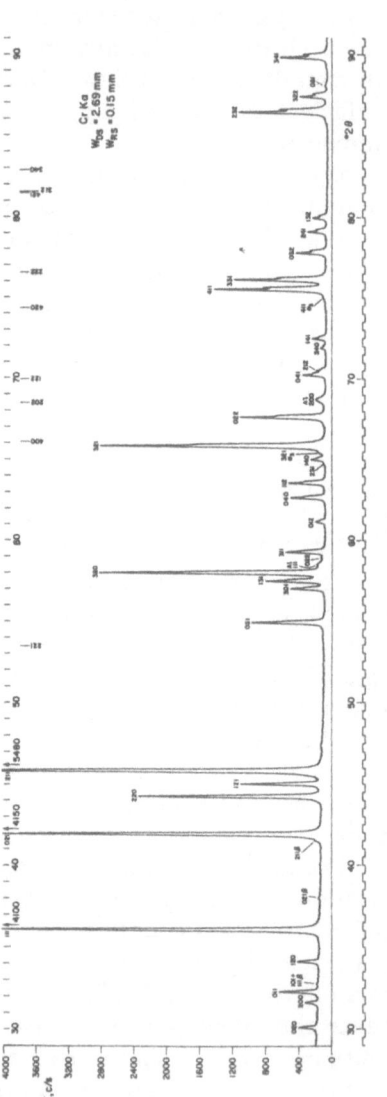

Fig. 10 (a)

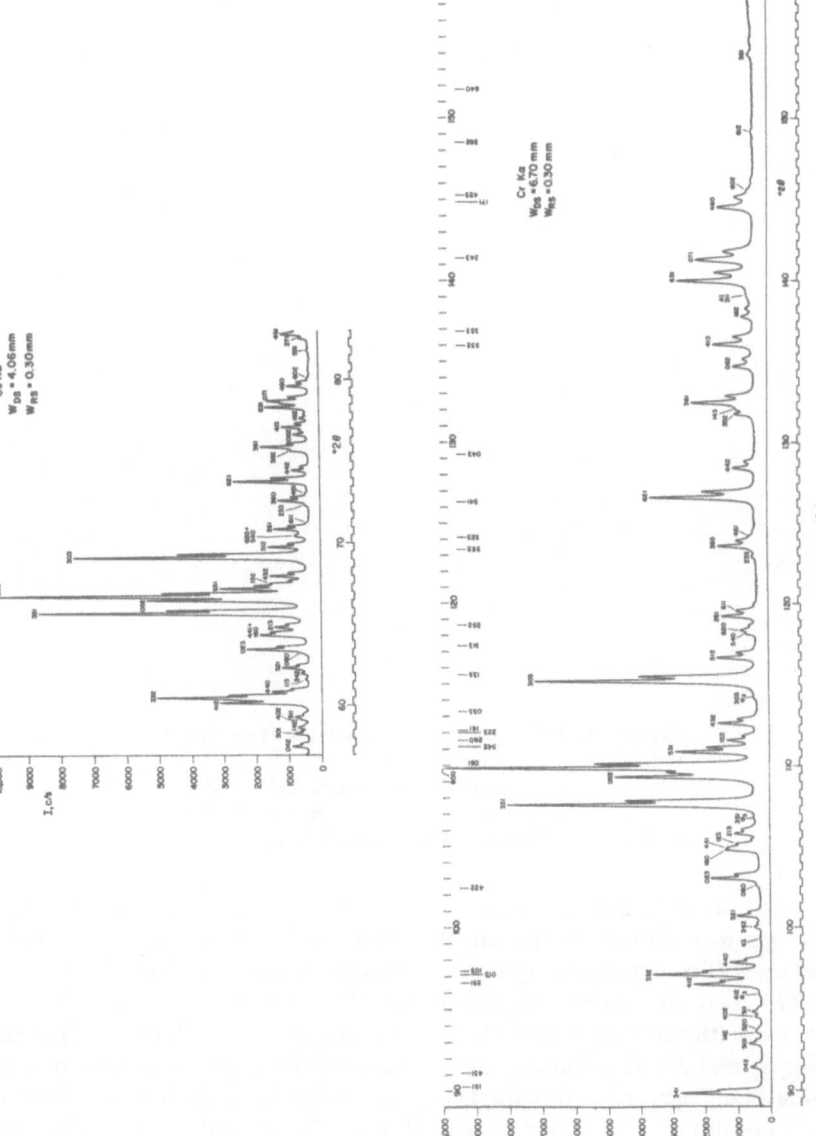

Fig. 10 (b)

Figure 10. Diffractometer patterns of topaz with Cu Kα and Cr Kα radiation. (a) d = 4.4 to 1.6 Å; (b) d = 1.6 to 1.16 Å.

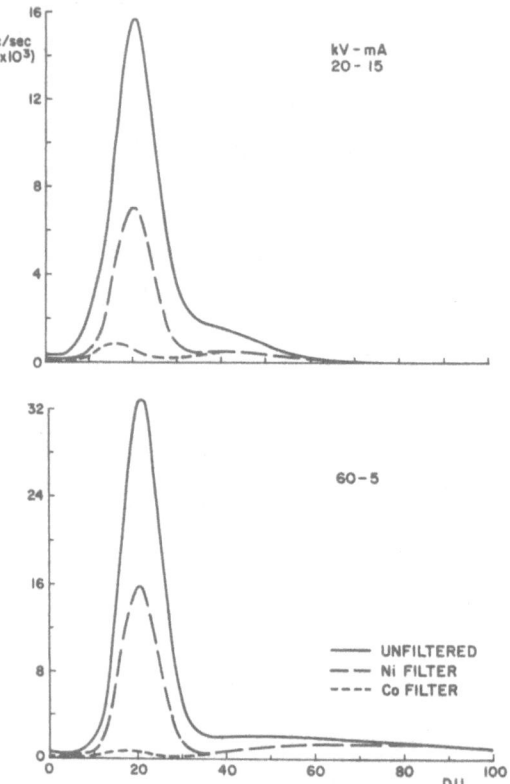

Figure 11. Pulse amplitude distributions of direct radiation from a copper anode X-ray tube, $\psi = 10^\circ$, 0.25-mm thick beryllium window, DC constant potential, and same total power. Balanced filters, NaI·Tl scintillation counter and 1 DU window.

Two methods may be used to derive the ratio. In the first method, applicable to unfiltered radiation, the total number of quanta of all pulse amplitudes (i.e., all wavelengths) is determined with the base level set at zero DU and the upper level removed. The number of counts is then measured with the analyzer window set to include nearly all ($\approx 96\%$) of the Cu $K\alpha_{1+2}$ and $K\beta$ distributions using settings for monochromatic radiation previously determined with a single crystal. If the $(K\alpha_{1+2}/K\beta)$ ratio is known, $(K\alpha_{1+2}/\text{continuous})$ can then be calculated. The second method gives the filtered ratio and is used without discrimination. The difference in the number of counts observed with balanced nickel and cobalt filters is taken as Cu $K\alpha_{1+2}$, and the counts with the cobalt filter is the same as the nickel-filtered continuous radiation with $K\alpha$ and $K\beta$ removed. This is justified if the filters are properly balanced, and in this case both filters gave the same number of counts above 40 DU. The nickel filter 0.02 mm thick reduced $K\alpha_{1+2}$ by a factor of 1.8 and practically eliminated $K\beta$ so that the distributions in figure 11 were narrowed and shifted slightly to lower pulse amplitudes. The cobalt filter caused a further slight shift and elimination of the peak of the characteristic lines.

The unfiltered ratio $(K\alpha_{1+2}/\text{continuous})$ at $60/40/20$ kV constant power was $3/4/6$ and for the filtered case was $2/3/4$. The ratios include the assumptions of $(K\alpha_{1+2}/K\beta) = 6$ and a 25% addition to the observed $K\alpha_{1+2}$ intensity to correct for the transmission of the air path, X-ray, and counter tube windows. Somewhat more precise values would be obtained by measuring the relative intensity of $K\alpha$ and $K\beta$ with a single crystal and correcting for Lorentz polarization and other factors and using a long evacuated cylinder with thin beryllium windows to eliminate most of the air path. The transmission correction was applied only to $K\alpha$ and is not entirely correct since the low intensity soft continuous radiation is attenuated even more. In any case it is clear that the primary beam produces more Cu $K\alpha$ quanta than the entire continuous radiation.

The spectrum of a copper target tube operated at 40 kV and recorded using a single crystal plate of silicon cut parallel to (111) is shown in figure 12. Introduction of a nickel $K\beta$-filter decreases the intensities of the lines and the continuous radiation (middle curve, figure 12). The spectral range detected may be restricted by an amount determined by the wavelength dependence of the quantum counting efficiency and resolution of the detector and the settings of the pulse height analyzer[1]. The combined use of the $K\beta$-filter and pulse amplitude discrimination provides a high degree of monochromatization, as shown in the lower curve. This electronic method causes a much smaller loss of $K\alpha$ intensity than that which occurs with a crystal monochromator. The specimen scatters the entire incident spectrum with varying efficiency, depending on the structure, wavelength, and angle. The specimen may also add to the spectrum by producing characteristic line X-ray fluorescence induced by that portion of the primary spectrum having wavelengths shorter than the absorption edges of the elements in the specimen.

3.3. Optimum Conditions for X-Ray Tube Operation

Although there are no sharply defined optimum conditions there are certain general principles which should be followed to achieve good results. The penetration of electrons in the anode depends on the applied voltage and atomic number of the anode element. When the voltage is increased

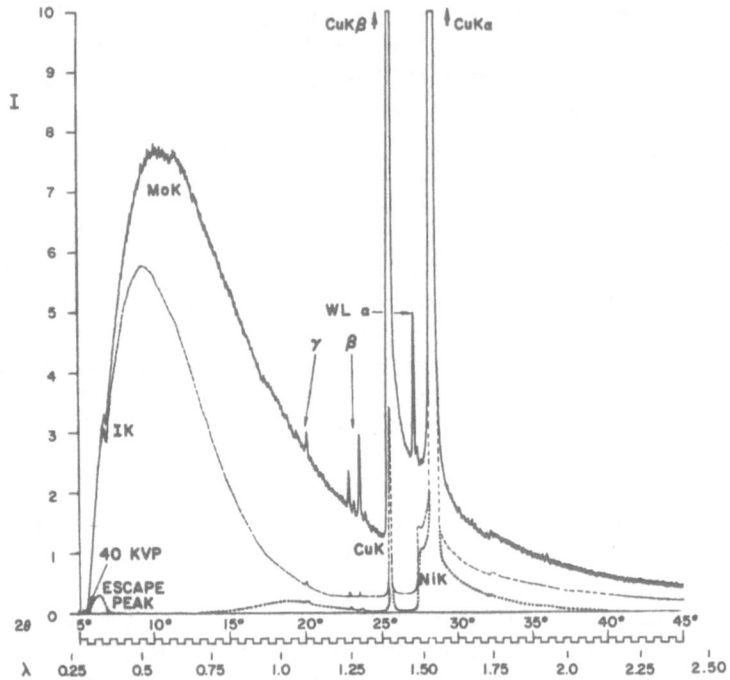

Figure 12. Upper curve: spectrum of copper anode X-ray tube, 40 kVp (full-wave·rectification), ψ 6°, silicon single crystal plate cut parallel to (111), NaI·Tl scintillation counter. Middle curve: same with 0.014-mm thick nickel filter. Lower curve: filter and pulse height analyzer set to transmit 90% of Cu $K\alpha$. The weak W L impurity lines are eliminated by the nickel filter.

X-rays are generated at greater average depths t and hence the absorption of the emerging rays increases. Since ψ is relatively small the self-absorption in the path length $t/\sin \psi$ may cause large losses of intensity. The effect is large enough in some cases to cause a reduction of intensity when the voltage is raised.

The dependence of intensity on kV and ψ has been studied by a number of authors using the electron microprobe[10-12] and diffraction tubes[13]. It may be measured using a powder or single crystal reflection and varying the kV using constant current (CC) or also varying the current to obtain constant power (CP); alternatively, the CC data may be multiplied by the appropriate power factor to convert to CP. Another method is to use the primary beam, balanced filters, and pulse amplitude discrimination, providing suitable small apertures are employed to limit the intensity to the linear range of the detector and circuits, as described in Section 3.2. The data are applicable to various tubes having the same window transmission and anode metal, providing the anode surfaces are flat and polished and the voltage waveform or ripple is the same.

A typical set of data is shown in figure 13 for W $L\alpha$ radiation, $\psi = 6°$. The intensity rises rapidly over a short voltage range above the critical excitation potential and then increases linearly and gradually falls off at higher voltages as the anode self-absorption increases. The voltage range and slope of the linear region depend on the anode, the wavelengths, and ψ. For example, the linear region using CC conditions for Cu $K\alpha$ increased from 23 to 32 kV at $\psi = 6°$ to 25 to 39 kV at $\psi = 10°$, and the maximum intensity for CP moved from 45 to 50 kV.

The dependence of Cu $K\alpha$ intensity on kV and ψ is plotted in two different forms in figure 14. These data were obtained from measurements of the direct beam using a specially designed device to vary ψ, and the X-ray tube window was mounted lower than normal in the manufacture to allow a ψ range of 0 to 20°. The upper graph shows the gradual decrease of intensity with increasing kV at small ψ's. The rate of increase rises with increasing ψ, and the 60-kV curve appears to be near its peak at $\psi = 15°$. The lower graph illustrates the small dependence of intensity on ψ when the kV is low, and by contrast the much larger dependence as the kV is increased. For example, at $\psi = 6°$ the intensity is nearly the same from 40 to 60 kV.

The application of such data to powder diffractometry is illustrated in figure 15 for Cu $K\alpha$ at $\psi = 16°$ and Cr $K\alpha$ at $\psi = 10°$ using the silicon

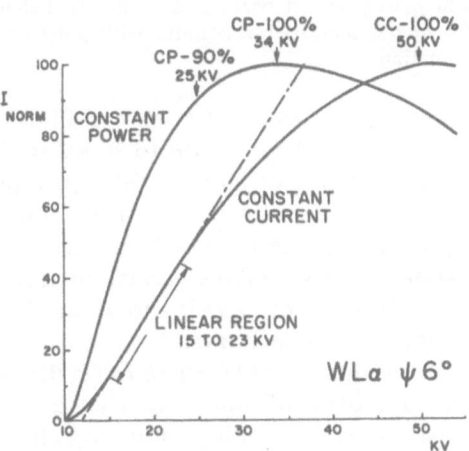

Figure 13. Dependence of W $L\alpha$ intensity on kV, ψ 6°, silicon (111) powder reflection. *CC* constant current; *CP* constant power.

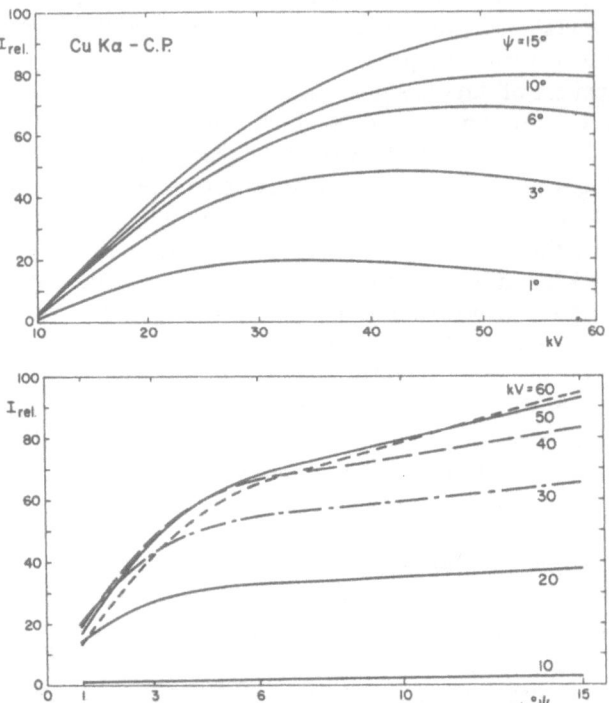

Figure 14. Dependence of Cu $K\alpha$ intensity on kV and ψ for constant power. Direct beam measurements, balanced Ni and Co filters, scintillation counter with pulse amplitude discrimination.

(111) reflection. The Cu $K\alpha$ intensity P-B increases linearly to 52 kV using CC and to 32 kV using CP and for Cr $K\alpha$ to 30 kV (CC) and 22 kV (CP). Above the linear range the rate of increase falls off rapidly, and although the Cu $K\alpha$ CP curve has not quite reached its peak at 60 kV, the Cr $K\alpha$ CP curve peaks at about 37 kV. The observed background increases linearly and is shown separated as wavelengths greater or less than Cu $K\alpha$. As a result of the different dependence of P-B and B on kV, the $(P$-$B)/B$ curve peaks at a lower voltage than the P-B curve.

The selection of criteria such as $(P$-$B)^2/B$ or $(P$-$B)^3/B$ gives more weight to the intensity, which is often of more importance than the background, and the peaks of such curves occur at higher voltages than the $(P$-$B)/B$ peak. If the Cr X-ray tube is rated at 1 kW and operated at 25 kV the current should be 40 mA for maximum power, or 80 mA for a 2-kW tube. Unfortunately, present-day diffraction tubes cannot be operated at such high

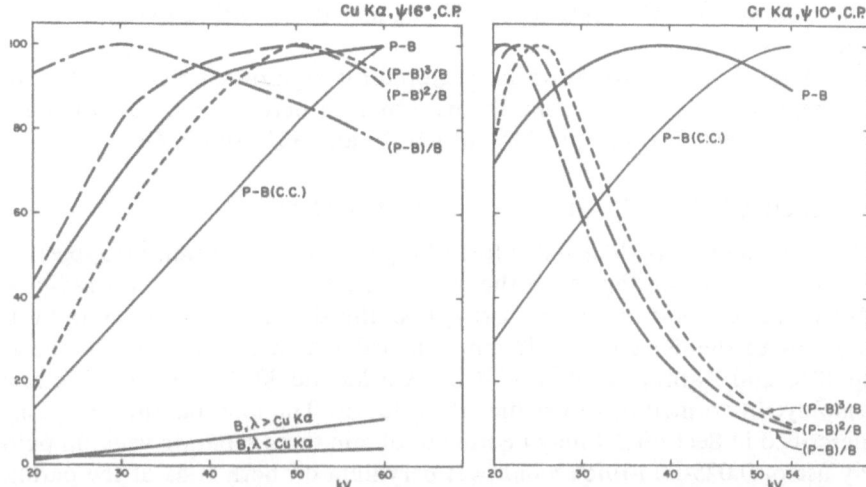

Figure 15. Variation of peak intensity *P-B* and other criteria on kV for Cu *K*α at $\psi = 16°$ and Cr *K*α at $\psi = 10°$. All curves for constant power except *P-B* curves labeled *CC*. Silicon (111) powder reflection, Ni or V β-filters, scintillation counter with pulse amplitude discrimination, vacuum diffractometer, fixed time measurements.

currents, and it will be necessary to develop tubes with much higher current capabilities. By increasing the angle of view the peaks of the curves are shifted to higher voltages, thereby keeping the required current to a minimum. Thus, in practice it is advisable to select the largest possible ψ consistent with the maximum required angular aperture α that can be transmitted by the window, making certain there is no cut-off of the focal line length. In some diffractometers it is necessary to remount the incident parallel slit assembly to increase ψ([14]).

The permissible electric power load of the X-ray tube is dependent on a number of factors, particularly thermal conductivity, melting point and physical properties of the anode, the focal line dimensions, the efficiency of heat transfer from the target surface to the cooling water, the temperature rise of the windows, and similar factors. Generally, the narrower the focal line the higher the specific loading (i.e., power per unit area), and the total load is usually proportional to the length. Although increasing ψ increases the projected width of the focus, thereby adding to the breadth of the profiles (Section 2.2), tubes with a narrow line focus approximately 0.4 × 8 mm may be used to minimize this effect. These tubes with Cu, Mo, and W anodes are conservatively rated at 1.2 kW at 50 kV or 375 W/mm²; similar tubes with Cr anodes are rated at about 1 kW. The popular 1.2 × 10 mm focus Cu tubes have about the same rating so that the specific load is 100 W/mm², and the wide-focus 3 × 12 mm tubes with 2 kW rating have only 56 W/mm².

Thin beryllium windows are required for high transmission of the soft radiations. In the experimental work reported here the window thickness was 250 μ, which transmits 93 % of Cu $K\alpha$ and 83 % of Cr $K\alpha$.

4. REDUCTION OF AIR-PATH ABSORPTION

The long air path causes a loss of intensity and may cause troublesome scattering at small angles. In the case of a diffractometer with a radius of 185 mm and a 30-mm-radius X-ray tube, the air path from the X-ray tube window to the detector is 373 mm. The calculated transmission of the air at 20°C and 1 atm pressure is 68 % for Cu $K\alpha$ and 30 % for Cr $K\alpha$. We have employed two methods to reduce these losses. The long parallel slit design described in Section 2.4 may be used to obtain three separate vacuum paths by use of 0.025-mm-thick windows: beryllium on both ends of the parallel slit boxes and Mylar on the slot opening of the cylindrical specimen chamber. The calculated transmission through the remaining 66 mm of air and six windows is 86 % for Cu $K\alpha$ and 64 % for Cr $K\alpha$.

Alternatively, a vacuum chamber of the type shown in figure 16 may be used([14]). This chamber rotates with the specimen, and the vacuum is pulled from the right side of the diffractometer through the central shaft of the goniometer. The remaining 117 mm and two 0.025-mm-thick Mylar windows transmit 84 % Cu $K\alpha$ and 58 % Cr $K\alpha$. Helium could be leaked into the chamber to avoid the vacuum. Both systems provide radiation protection since most of the X-ray path is enclosed.

5. STEROID PATTERNS

The steroids provide a good example of the advantages of the techniques described in this paper. They are an important family of substances in biology and medicine, and a number of the crystal structures have been determined from single crystal studies. Parsons and his co-workers([15,16]) have published powder pattern data for 502 steroids obtained with a 114.6-mm-diameter Debye–Scherrer powder camera and Cu $K\alpha$ radiation; the intensities were obtained with an automatic recording microdensitometer. He supplied a number of steroids so that the results could be compared with diffractometer data. The steroids give excellent powder patterns with sharp high-intensity lines. Although most of the powder patterns are complex and cannot be used for crystal structure determination, they are useful for identification purposes. No attempt was made to index the patterns by direct methods, and single crystals were not available.

Figure 16. Vacuum chamber mounted on Norelco diffractometer showing rayproof slit mount on X-ray tube tower, access window, and scintillation counter.

Patterns of the four steroids shown in this paper are listed below, and Parsons' papers([15]) should be consulted for further details.

$$C_{30}H_{50}O: \text{Parsons IV, pattern No. 24}$$

$$C_{28}H_{46}O_2: \text{Parsons III, pattern No. 5}$$

$$C_{23}H_{31}O_4N: \text{Parsons IV, pattern No. 58}$$

$$C_{18}H_{22}O_2: \text{Parsons IV, pattern No. 99}$$

The experimental conditions employed were as follows: vacuum diffractometer, radius 185 mm, Cr anode X-ray tube, focal line 0.07 × 8 mm at $\psi = 10°$, 30 kV, 20 mA, 0.012 mm V β-filter, scintillation counter with pulse amplitude discrimination, scanning speed 0.5 (2θ)/min, time constant 4 sec, receiving slit 0.05°, two short parallel slit assemblies each with $\delta = 4.5°$, incident assembly axial length 12 mm, and receiving assembly 20 mm. The divergence slit width was reduced at 30° to permit scanning to small angles, and the intensities below this angle should be multiplied by 3.88 to put all reflections on the same scale. In the range 60° to 30° $\alpha = 1°$, antiscatter slit 0.75 mm; 30° to 3°, $\alpha = 0.25°$, antiscatter slit 0.4 mm. It is essential to use small divergence slit widths and carefully center the beam at the smallest scanning angle when scanning in the small-θ region. It is also necessary to reduce the X-ray tube voltage to about 30 kV to avoid difficulties with escape peaks at small angles[17]. Flat rotating specimens were used.

Figure 17 shows a pair of recordings of $C_{30}H_{50}O$ obtained with Cu $K\alpha$ and Cr $K\alpha$ radiations using the same experimental conditions and specimen. The reflections common to both patterns are indicated by small vertical lines. The small circles above some of the lines indicate those reflections found in the Cr pattern which were not resolved in the Cu pattern. The Cr pattern has 80 reflections in the d range 33.3 to 2.35 Å and the Cu has only 44; the missing reflections were all of relatively low intensity. The increased dispersion causes a larger separation of the pair of strong lines at 12° in the Cu pattern and the relative intensities are changed from 86 : 100 to 64 : 100.

Considerable care was taken in the specimen preparation to minimize preferred orientation, but some may still be present. The high peak at 38° in the Cu pattern may be caused by a relatively large crystal near the specimen surface which was not reached by the less penetrating Cr radiation. Table I compares data from the Cr diffractometer pattern with Parsons' Cu film for reflections down to 5 Å. The writer also obtained powder camera films with Cu radiation, and these checked with Parsons' data. However, only 26 reflections were found, and a number of these (indicated by an asterisk) were measured in the Cu diffractometer pattern. The most important missing reflection is the weak peak at 33.3 Å. These large differences in the data from the same specimen are due to the much poorer resolution and dispersion, the inaccessible low-angle region, and the difficulties of obtaining good intensity data from the film.

Figure 18 shows three additional recordings of steroids obtained with Cr $K\alpha$ radiation, and table II compares the three most intense lines with those obtained by Parsons. The agreement is poor for the more complex patterns and improves as the patterns become more simple.

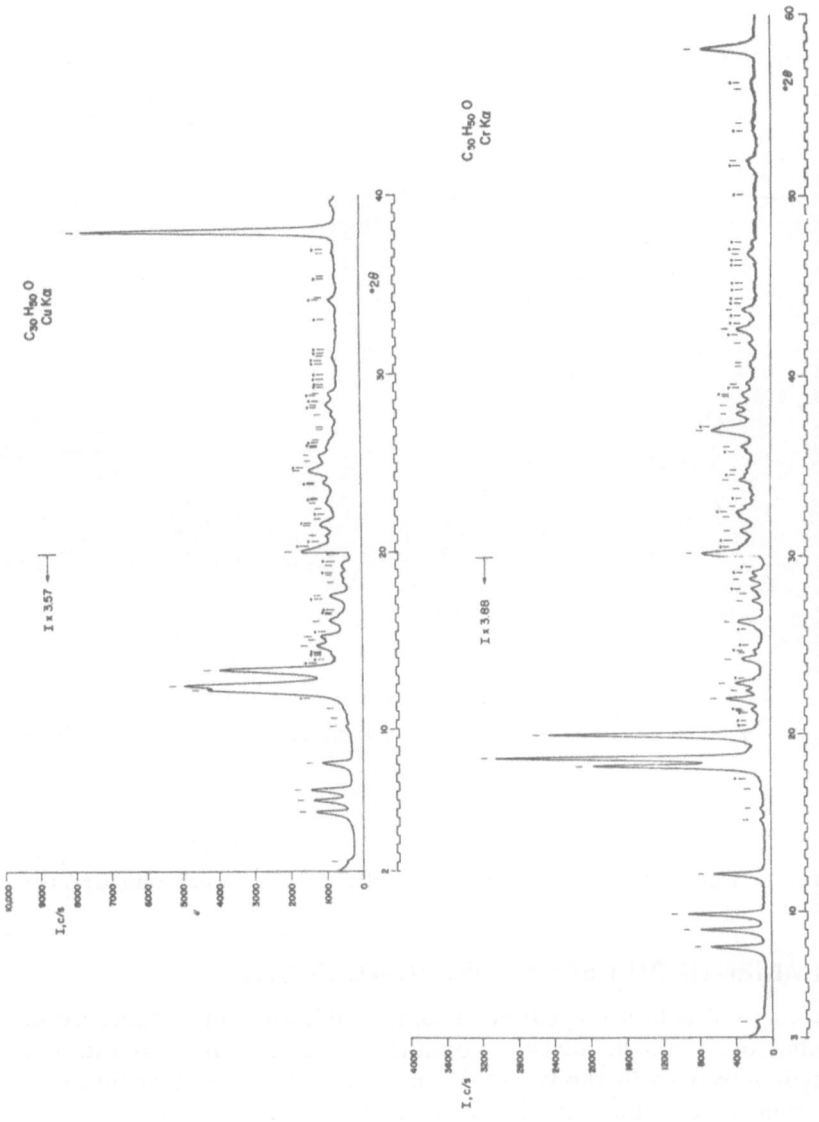

Figure 17. Diffractometer recordings with Cu $K\alpha$ radiation (above) and Cr $K\alpha$ (below). The small solid circles indicate reflections identified in the Cr pattern but could not be found in the Cu pattern. Experimental conditions described in text.

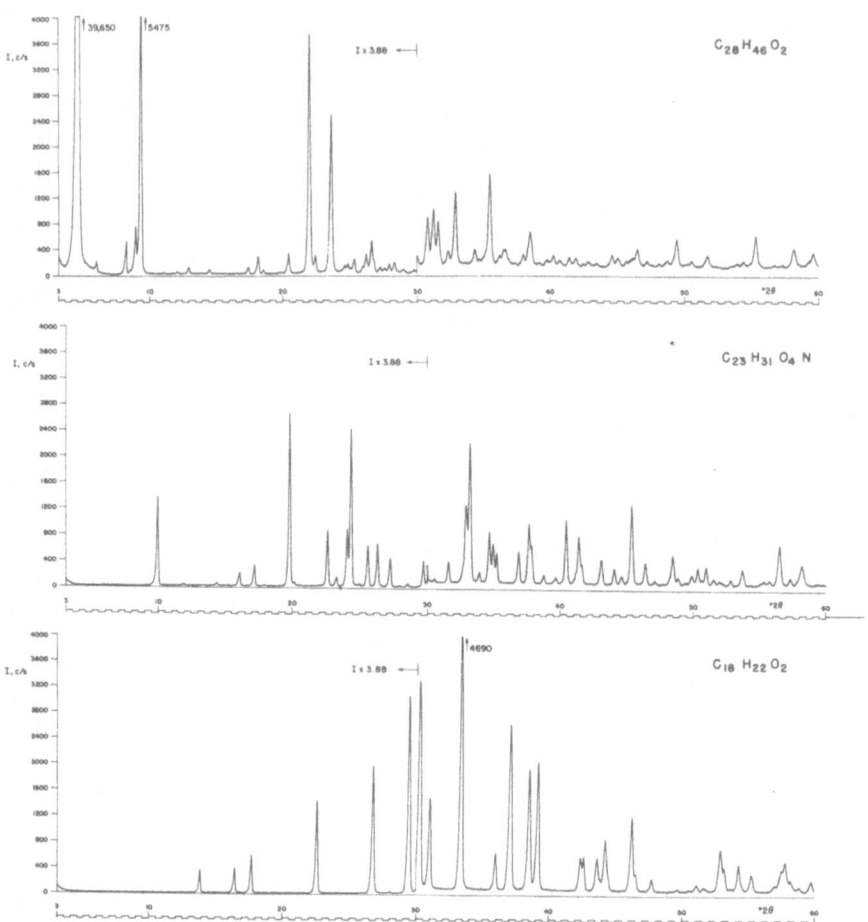

Figure 18. Typical diffractometer recordings of steroids obtained with Cr $K\alpha$ radiation.

6. COMMENTS ON RECORDING POWDER DATA

Powder diffractometry requires a considerable amount of data readout and reduction. With the advent of on-line computers it is likely that new and vastly improved methods will become available in the near future. In this section some factors are discussed which are pertinent to widely used existing methods.

The measurement of a strip chart recording of a complex pattern inevitably involves a certain amount of subjectivity in the selection of the experimental conditions and the data measurements. Two quantities are

Table I. Comparison of Data for $C_{30}H_{50}O$ Obtained with Diffractometer and Powder Camera

d		I_{rel}		d		I_{rel}	
WP[a]	JP[b]	WP	JP	WP	JP	WP	JP
33.3		3		6.20*[c]		2	
16.1	15.81	20	10	6.17	6.18	3	18
14.4		24		6.14*		3	
13.1	13.11	29	15	5.98		14	
10.8	10.80	19	20	5.87		5	
8.60		1		5.76	5.78	11	100
8.26		1		5.70*		3	
7.73		1		5.44	5.44	7	27
7.49*		1		5.36		1	
7.20	7.17	64	50	5.33*		3	
7.04		100		5.27*		1	
6.57	6.64	80	32	5.09*	5.06	3	37
6.41*		1		5.02		9	
6.33*		2					

[a] WP—Diffractometer data with Cr $K\alpha$.
[b] JP—Parsons' camera data with Cu $K\alpha$.
[c] Asterisk indicates line also observed with Cu $K\alpha$.

required, intensity and d spacing. The d's are derived from the 2θ's and the precision in reading the angles is dependent largely on the counting statistical accuracy of the intensity measurements. No special difficulties arise when the lines have a good peak-to-background ratio, are resolved, and are spaced relatively far apart. Locating small peaks in the background involves counting statistical factors which in practice may not easily be applied, particularly to rate meter recordings. As the lines occur closer together the tails of the profiles overlap and raise the apparent background in the immediate vicinity. The peak intensities and hence the 2θ-angles corresponding to those peaks are then modified. The intensities of small peaks near the tails of high peaks must be estimated by sketching the continuation of the high peak tail under the small peak. In very crowded regions the individual profiles never reach their true background, and the peak heights can only be estimated. The shape of the broadened composite profiles is often the most important key to the number of overlapping reflections. The necessity of using the best possible experimental conditions to achieve maximum resolution, dispersion, and counting statistical accuracy soon becomes evident. The possibility of applying deconvolution methods to sharpen the profiles with a computer has not been attempted in powder

Table II. Comparison of Three Highest Intensity Lines of Steroids
Obtained with Diffractometer and Debye–Scherrer Camera

	(WP) Diffractometer-Cr $K\alpha$				(JP) D–S Film-Cu $K\alpha$		
$C_{30}H_{50}O$							
d	7.04	6.58	7.20		5.78	7.17	5.06
I_{rol}	100	80	64		100	50	37
(JP)		32	50	(WP)	11	64	3
					Missing reflections: 54		
$C_{28}H_{46}O_2$							
d	28.4	13.9	5.97		5.89	5.61	5.04
I_{rel}	100	14	9		100	67	49
(JP)			100	(WP)	9	6	4
					Missing reflections: 47		
$C_{23}H_{31}O_4N$							
d	6.62	5.40	5.47		5.40	5.79	4.99
I_{rel}	100	91	33		100	88	63
(JP)	50	100		(WP)	91	32	25
					Missing reflections: 31		
$C_{18}H_{22}O_2$							
d	4.50	4.93	5.81		4.25	4.93	5.83
I_{rel}	100	65	47		100	75	56
					Missing reflections: 12		

diffractometry because it requires a knowledge of all the complete profile aberrations; in any case the success would depend on the quality of the original data.

The most common and useful method of recording powder patterns is with a rate meter and strip-chart recorder. This method assumes an exactly uniform relationship among the chart movement, goniometer scanning speed, and registration of the intensity data. The charts are subject to at least three possible sources of error: (1) uneven movement of the chart, (2) limited precision in reading the chart, and (3) time constant and scanning speed distortions.

A number of commercial recorders were tested, and none had a uniform chart drive. When accurate time markers were recorded the linear distance between the marks often varied ± 0.005 to 0.010 in. per inch of chart, and these variations were usually random. The resulting angular error depends on the chart and goniometer speeds, and one cannot be certain if the variations were spread over the entire distance between the marks or occurred as a sudden jump. The recorder paper dimensions may change with humidity, and the precision of the printed grid often varies. The degree marker on the diffractometer must also be adjusted to give accurately spaced signals to the recorder.

Commercial instruments are usually available with four chart speeds on the recorders, $7\frac{1}{2}$, 15, 30, and 60 in./hr, and with five diffractometer speeds, $\frac{1}{8}$, $\frac{1}{4}$, $\frac{1}{2}$, 1, and 2° (2θ)/min. These allow many combinations or recording conditions and the two extremes are

Goniometer speed, °2θ/min	$\frac{1}{8}$	2
Time to scan 80°, hr	10.8	0.66
Chart speed, in./hr	60	7.5
80° chart length, ft	54.1	0.4
1 in. on chart, deg	0.125	16
±0.005 in. on chart, deg	0.0006	0.08

It is evident that neither of these extremes is practical, and the selection of a particular combination depends on such factors as the complexity of the pattern and the precision required for the analysis.

It is well known that increasing the product of the scanning speed V and time constant of the rate meter T (including the recorder pen response) decreases the observed peak intensity, broadens the profile asymmetrically, and shifts the peak position in the direction of the scan[18]. The line profile shape and the other changes are the same for various combinations of V and T, providing their product is the same. For a given set of experimental conditions the counting statistical accuracy decreases with increasing V regardless of T. If V and T are both large the distortions will be large and serious errors will occur in complex patterns. The magnitude of the distortion is also dependent on the slopes of the profiles and is different when scanning a partially resolved $K\alpha$ doublet or closely spaced pair of reflections from opposite directions. The distortion is thus dependent on the line shape, resolution, and receiving slit width.

Figure 19 shows the decrease of peak intensity ΔP and peak shift $\Delta 2\theta$ are not exactly the same for reflections with different line shapes when VT is large. The data are for the same receiving slit aperture and the recordings were toward decreasing 2θ so that $\Delta 2\theta$ is minus. For $VT = 0.1$ to 1, $\Delta 2\theta \leqslant 0.01°$ and $\Delta P \leqslant 7\%$. Neither $\Delta 2\theta$ nor ΔP extrapolate linearly to zero from large values of VT, although the extrapolation is virtually linear for $VT < 2$. Klug and Alexander[19] state that T should be less than the time width of the receiving slit which they define as the time required for the slit to traverse its own width. For example, if $\epsilon_{RS} = 0.08°$ and $V = 0.25°$/min, T should be less than 1.9 sec.

The precision in determining the peak angles from a rate meter chart

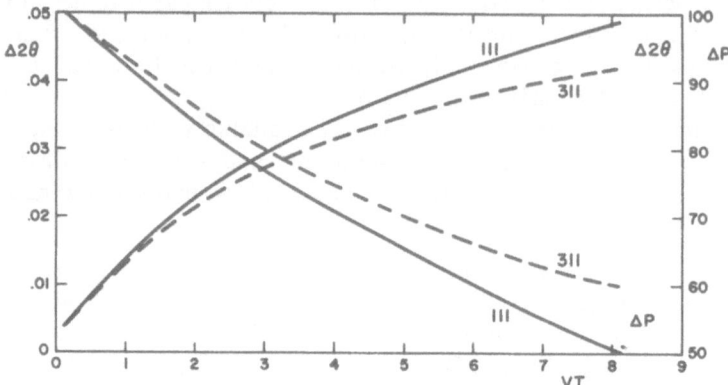

Figure 19. Decrease of peak intensity ΔP and shift of peak position $\Delta 2\theta$ as a function of the product of the diffractometer scanning speed V and time constant of the rate meter and recorder T. Silicon powder specimen (111) Cr $K\bar{\alpha}$ and (311) Cr $K\alpha_1$, ϵ_{RS} 0.04°, scanned from high to low 2θ.

is illustrated in table III. The chart speed was varied and $VT \leqslant 1$ in all cases. Six peaks of various intensities and widths were run twice. Two observers measured each profile twice, and their agreement was within the precision of the measurements. The chart-measuring device has a large 3 × magnifier, and the scale was graduated to 0.020 in.; the position of the cursor could be estimated reproducibly to $\frac{1}{4}$ division. The degree marks on the charts were used as reference points. The uncertainty ϵ that was possible in measuring the peak angles was determined by varying the position of the cursor to the maximum reasonable missetting decided by visual observation of the uppermost portion of the profile. It is clear that there was some subjective bias in determining ϵ. The precision in setting the cursor on the peaks was about the same as the precision in reading the scale for the strong narrow peaks (111), (321), and (531) and was reduced by about a factor of two for the weak and broader peaks (140), (162), and (631). There was no gain in precision in using a very long chart (1° = 4 in.), and indeed the weak and broad lines become more difficult to measure. The short chart (1° = $\frac{1}{2}$ in.) limits the precision, and the best compromise appears to be 1° = 1 or 2 in.

   The most precise method of locating peak diffraction angles is by step scanning across the peak in equal angular increments, measuring the intensities for a fixed time, and finding the maximum of a parabola fitted to the three highest intensities. This method eliminates most of the difficulties described above for the rate meter method, and practically all subjectivity is avoided aside from the selection of the angular increment

Table III. Dependence of Precision on Recording Conditions[a]

					hkl	111ā	140α₁	321α₁	351α₁	162α₁	631α₁
				approximately °2θ		36	65	66	108	138	140
				P, c/s		3000	100	1800	4300	200	1100
				$W_{1/2}$, °2θ		0.13	0.19	0.19	0.19	0.21	0.23
V °2θ/min	VT sec °2θ/min	Chart speed in./hr	Chart length in./°2θ	Scale accuracy ±°2θ			±ε, °2θ				
$\frac{1}{8}$	$\frac{1}{4}$	30	4	0.0013		0.003	0.003	0.003	0.003	0.009	0.006
		15	2	0.0025		0.003	0.006	0.003	0.003	0.013	0.006
		7.5	1	0.005		0.005	0.013	0.005	0.005		
$\frac{1}{4}$	$\frac{1}{2}$	30	2	0.0025		0.003	0.006	0.003	0.005	0.013	0.008
		15	1	0.005		0.005	0.013	0.005	0.005	0.013	0.010
		7.5	½	0.01		0.01	0.01	0.01	0.01	0.02	0.01
$\frac{1}{2}$	1	60	2	0.0025		0.003	0.006	0.003	0.005	0.013	0.008
		30	1	0.005		0.005	0.010	0.005	0.005	0.010	0.010
		15	½	0.01		0.01	0.01	0.01	0.01	0.02	0.01
1	1	60	1	0.005		0.005	0.010	0.005	0.010	0.015	0.010
		30	½	0.01		0.01	0.01	0.01	0.01	0.02	0.01

[a] Curved topaz powder specimen, Cr Kα radiation, vacuum diffractometer.

and counting time. If the counting statistics are poor, the precision may be increased by increasing the counting time, and if the peak is broad the angular increment may be increased.

The intensities I_A, I_B, and I_C are measured for equally spaced angles $2\theta_A$, $2\theta_B$, and $2\theta_C$, respectively, where I_B is the experimental point with the highest intensity. The maximum $2\theta_P$ of a fitted three-point parabola is found from the expression

$$2\theta_P = 2\theta_B + \frac{\delta}{2}\left[\frac{(I_B - I_A) - (I_B - I_C)}{(I_B - I_A) + (I_B - I_C)}\right] \tag{12}$$

where $\delta = 2\theta_C - 2\theta_B = 2\theta_B - 2\theta_A = 0.01$ to $0.02°$. The precision may be increased by repeating the calculation for three points separated by twice the angular increment using the uppermost five or six intensities and averaging the answers. The calculation takes a few minutes using a slide rule or desk calculator, and if a large number is to be made the data may be punched directly on paper tape and fed to a computer.

The results were compared with those obtained using Wilson's method([20]), in which a least-squares parabola is fitted to a range of observations evenly distributed over the peak. The three-point parabola calculation is easier to make, and the precision is the same. The reproducibility of the parabola method was checked with a large number of reflections whose intensities were purposely varied by a factor of six to change the counting statistical accuracy, and breadths were varied by a factor of two or more. The average reproducibility was $\pm 0.002°$ (2θ). It is evident that the calculated peak location of asymmetric profiles depends on the range of observations chosen, and hence it is essential to select a range near the peak.

The centroid measurement is not possible in complex patterns because of the overlapping, and the midpoint of a chord[1] at various heights above background will also be influenced by line profile asymmetry and overlapping. Therefore, the angle corresponding to the peak appears to be the only feasible measure applicable to complex patterns.

ACKNOWLEDGMENTS

The writer is indebted to his former colleagues Miss Marian Mack and Mrs. J. Taylor who collaborated in various phases of the program. Mr. I. Vajda was responsible for much of the engineering design and Mrs. P. Harnack aided in data collection and reduction.

REFERENCES

1. W. Parrish (ed.), *X-Ray Analysis Papers*, Centrex Publishing Co. Eindhoven (1965).
2. A. J. C. Wilson, *Mathematical Theory of X-Ray Powder Diffractometry*, Philips Technical Library, Eindhoven (1963).
3. R. E. Ogilvie, Parafocusing diffractometry, *Rev. Sci. Instr.* **34**, 1344–1347 (1963).
4. W. Parrish, M. Mack, and J. Taylor, Determination of apertures in the focusing plane of X-ray powder diffractometers, *J. Sci. Instr.* **43**, 623–628 (1966).
5. W. Parrish, M. Mack, and I. Vajda, Seemann–Bohlin linkage for Norelco X-ray diffractometer, *Norelco Reptr.* **14**, 56–59 (1967).
6. W. Parrish and M. Mack, Seemann–Bohlin X-ray diffractometry: I and II, *Acta Cryst.* (in press).
7. W. C. Stoecker and J. W. Starbuck, Effect of Soller slits on X-ray intensity in a modern diffractometer, *Rev. Sci. Instr.* **36**, 1593–1598 (1965).
8. W. Parrish, Improved slits for X-ray powder diffractometers, *Rev. Sci. Instr.* **37**, 1607–1608 (1966).
9. L. S. Birks, R. E. Seebold, B. K. Grant, and J. S. Grosso, X-ray yield and line/background ratios for electron excitation, *J. Appl. Phys.* **36**, 699–702 (1965).
10. D. B. Brown and R. E. Ogilvie, Efficiency of production of characteristic x radiation from pure elements bombarded by electrons, *J. Appl. Phys.* **35**, 309–314 (1964).
11. M. Green, The angular distribution of characteristic x radiation and its origin within a solid target, *Proc. Phys. Soc.* **83**, 435–451 (1964).
12. R. E. Ogilvie, "Quantitative electron microprobe analysis," in *Proc. Eastern Anal. Symp.* (W. Parrish and H. van Olphen, eds.), Vol. 1, *X-Ray and Electron Methods of Analysis* (1967).

13. W. Parrish and N. Spielberg, Dependence of intensity on X-ray tube voltage, Joint Meeting Am. Cryst. Assoc. and Mineral Soc. Am., Montana State College, Bozeman (July 26–31, 1964).
14. W. Parrish, Vacuum chamber for Norelco diffractometer, *Norelco Reptr.* (in press).
15. J. Parsons, S. T. Wong, W. T. Beher, and G. D. Baker, X-ray diffraction powder data for steroids, *Henry Ford Hosp. Med. Bull.* Suppl. III, **11**, 23–52 (1963); Suppl. IV, **12**, 87–120 (1964).
16. J. Parsons, J. B. Holcomb, and W. T. Beher, X-ray diffraction powder data for steroids: *Henry Ford Hosp. Med. Bull.* Suppl. VIII, **15**, 133–138 (1967).
17. W. Parrish, Escape peak interferences in X-ray powder diffractometry, *in Advances in X-ray Analysis* (W. M. Mueller, G. R. Mallett, and M. J. Fay, eds.) Vol. 8, pp. 118–133, Plenum Press, New York (1965).
18. W. Parrish, X-ray intensity measurements with counter tubes, *Philips Tech. Rev.* **17**, 206–221 (1956).
19. H. P. Klug and L. E. Alexander, *X-Ray Diffraction Procedures*, John Wiley, New York (1954).
20. A. J. C. Wilson, The location of peaks, *Brit. J. Appl. Phys.* **16**, 665–674 (1965).

II. ENERGY DISPERSION X-RAY ANALYSIS USING RADIOACTIVE SOURCES

William J. Campbell

U.S. Department of the Interior, Bureau of Mines, College Park Metallurgy Research Center, College Park, Maryland

Radioisotope sources coupled with nondispersive techniques offer the following advantages over conventional dispersive systems: simplicity of design and low cost, compact lightweight instrumentation, and wide choice of excitation sources.

There are three types of radioisotope sources—α-, β-, and γ-emitters. Typical α- and β-emitters are Po^{210} and H^3/Zr, respectively. Generally beta emitters are used to generate primary X-rays which, in turn, excite secondary X-rays in the sample. Gamma sources may be either monochromatic, e.g., Fe^{55} (Mn K) or high-energy γ-emitters used in conjunction with a secondary emitter Am^{241}-Cs.

In nondispersive analysis, energy discrimination is achieved by selective filtration, balanced filters, differential absorption in the detector, and electronic pulse amplitude discrimination. All of these methods may be used individually or in various combinations.

Applications of these portable analyzers include prospecting and mining operations, monitoring of metallurgical processes, and automatic sorting of mail.

1. INTRODUCTION

During the early 1950's radioactive isotopes were considered as an excitation source to replace the X-ray tube used in fluorescent X-ray spectrography. However, replacement of the X-ray tube was not feasible because the low X-ray flux of the isotope sources was only about one-millionth of that obtainable from an X-ray tube. Nondispersive techniques were not feasible at that time because detector technology was limited to Geiger–Mueller counters. About ten years ago detectors became commercially available with dead times less than 1 msec and with output pulses proportional to the energy of the incident X-ray photon[1]. However, there was little progress in energy dispersion techniques for X-ray analysis until a few years ago.

During the past two years there has been a very significant increase in instrumentation and applications of X-ray spectrography using isotopic X-ray sources combined with energy dispersion by either electronic pulse-amplitude discrimination or selective X-ray filters. Most of the progress on applications is attributed to European and Japanese researchers who have exploited the economic and technological advantages of this technique. In the United States major improvements in low-energy isotopic source design and utilization have been achieved by research supported by the Division of Isotope Development of the Atomic Energy Commission. Major symposia on uses of low-energy X-ray and γ sources, including instrumentation and applications of radioisotope X-ray spectrography, were held recently[2-4] and an excellent general review has been published by Rhodes[5].

The purpose of this paper is to evaluate the present state of X-ray spectrography based on energy dispersion used in conjunction with isotopic X-ray excitation sources. Subjects to be considered include excitation, dispersion, and detection systems, a survey of some of the instruments that have been developed, and a brief discussion of present and projected applications.

2. INSTRUMENTATION

In conventional X-ray spectrography using an X-ray tube as the excitation source, the sample is subjected to a very intense X-ray flux of approximately 10^{13} photons/sec. The resultant characteristic X-rays are dispersed by means of X-ray diffraction and counted by a wide variety of detectors (figure 1). This wavelength dispersion system results in excellent

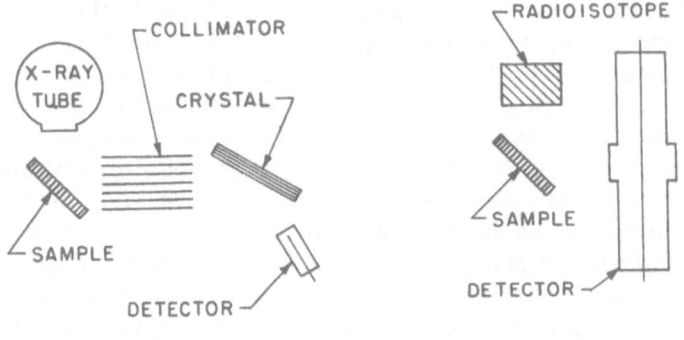

WAVELENGTH DISPERSION   ENERGY DISPERSION

Figure 1. Comparison of conventional X-ray spectrographic optics and energy dispersion[8].

spectral resolution, but the geometrical losses and low-diffraction efficiency reduce the X-ray intensity by a factor of more than one million. These wavelength dispersive systems require a stable 2 to 3 kW power supply, a precision goniometer, and a rather complex electronic readout. The cost of these spectrographs is in the range of $15,000 to $20,000 and upward.

Energy dispersion systems combined with a radioisotope X-ray source can be very elementary, as shown in figure 1. This simple system may be augmented by filters placed either between the source and the sample or between the sample and the detector. The detector is either a gas-filled proportional counter, a scintillation counter, or one of the recently developed lithium-drifted silicon or germanium detectors. In its simplest form this energy dispersion system may cost $2000 to $3000; with solid state detection systems the cost will be increased by a least a factor of 2.

2.1. Excitation Sources

We will consider first the radioisotopic excitation source consisting of α-, β-, or γ-emitters. Some commonly used low-energy X-ray and γ ray sources are listed in table I, taken from the paper by Rhodes[5]. The source activities range from about 1 mCi to 1 Ci. The number of photons generated per disintegration range from nearly unity to 1×10^{-5}. In general the X-ray flux from these sources are in the order of 10^7 photons/sec; the higher activity sources have a much lower photon-per-disintegration ratio. The cost of the source varies with the isotope and its activity. Excellent Fe^{55} sources are available for $50 to $100, whereas Cd^{109} costs about $250 to $300/mCi.

α-sources such as Po^{210} and Cm^{242} are used to excite very low-energy X-rays[6]. The principal advantage of an α-source is the high peak-to-background ratio, since the α-produced bremsstrahlung radiation is reduced by a factor of the square of the mass of the electron to that of the α-particle $(m_{electron}/M_\alpha)^2$. α-sources are a potential health hazard and must be handled with much greater care than the β- or γ-sources.

β-emitters, such as Pm^{147}, are used to generate bremsstrahlung and characteristic X-rays that serve as the source of excitation for the sample. These sources may be prepared by depositing a thin layer of the β-emitter on a suitable target material (apposite) or by mechanically or chemically mixing isotope and target. There is available an excellent compilation by Preuss[7] of the properties of three β-sources, Pm^{147}, Ca^{45}, and P^{32}.

Sources such as Fe^{55} which decay by K-electron capture are essentially monoenergetic—i.e., Fe^{55} emits manganese K X-rays. These sources are used to excite specific X-ray lines.

Table I. Commonly Used Low-Energy X-Ray and γ-ray Sources[5]

Source	Half-life	Useful radiations	Practical emission efficiency, (photons/disintegration)	Typical activity	Highest atomic number usefully excited, K X-rays
Iron-55	2.7 yr	manganese K X-rays, 5.9 keV	0.15	2 mCi	24
Tritium-zirconium	12.3 yr	bremsstrahlung, 2 to 12 keV	4×10^{-5}	units of 1 to 3 Ci	30
Cadmium-109	1.3 yr	zirconium L X-rays, 2 keV	10^{-5} to 10^{-4}	1 mCi	43
		silver K X-rays, 22 keV	0.8		
		γ-ray, 88 keV	0.04		
Promethium-147-aluminum	2.6 yr	bremsstrahlung, 10 to 100 keV	2×10^{-3}	0.5 Ci	60
Americium-241	470 yr	γ-ray, 59.6 keV	0.35	1 mCi	69
		γ-ray, 26 keV	0.02		
		neptunium L X-rays, 11 to 22 keV	0 to 0.2		
Gadolinium-153	236 days	γ-ray, 103 keV	0.2	1 mCi	88
		γ-ray, 97 keV	0.2		
		europium K X-rays, 42 keV			
Cobalt-57	270 days	γ-ray, 136 keV	0.10	0.5 mCi	98
		γ-ray, 122 keV	0.88		
		γ-ray, 14 keV	0 to 0.06		
		iron K X-rays, 6.4 keV			

2.2. Source–Sample–Detector Geometry

Two source–sample–detector geometries used with β-, γ-, and X-ray emitting isotopes are shown in figure 2. The central source arrangement is the one most widely used in radioisotope X-ray analysis. The important parameters are the sample–source–detector distances and the relative sizes of the three components. The X-ray filters are usually placed between the source and the detector window. This is the geometry used in the analyzer developed commercially by Hilger and Watts. Using the central source geometry, overall efficiency is 10^{-2} to 10^{-4} so that counting rates of 10^3 to 10^5 counts/sec are obtained from pure elements. Low output from the isotopic source makes shielding a minor problem, since the radiation hazard is not serious. Adequate shielding is provided by a shutter and by the sample

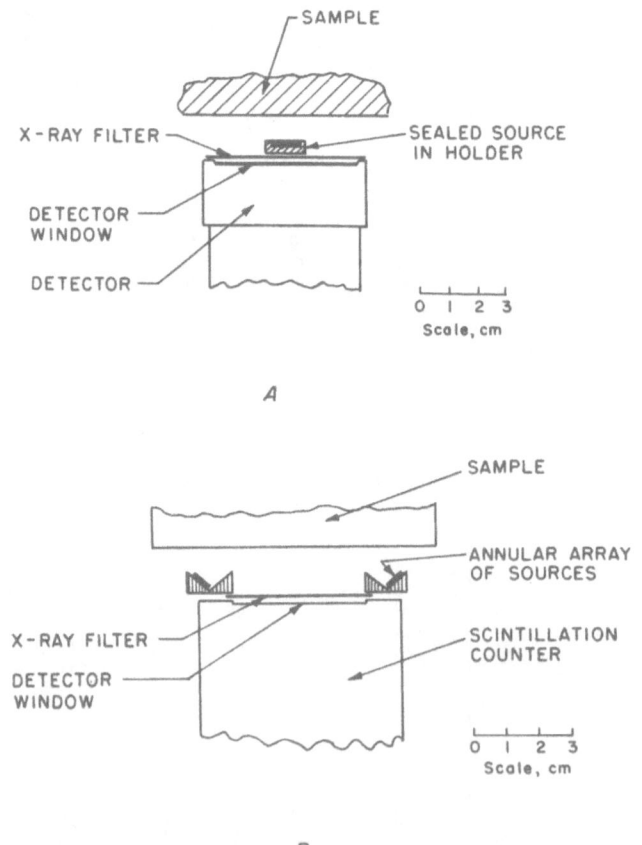

Figure 2. Central (A) and annular source (B) geometries[5].

being analyzed. At a distance of 1 ft away from the unshielded source, the radiation is only 1 mr/hr.

Another useful arrangement is the annular source (figure 2b), in which the sources are arranged in a doughnut-like array around the outside of the detector window. This type of source arrangement is necessary to prevent the source from blocking excited radiation from the sample reaching the detector. Our research group plans to use the new solid state lithium-drifted silicon detector with an annular source.

Source-target assemblies (figure 3) are used with β- and γ-emitters to give a characteristic spectra of the target element. This target element is selected so that the energy of its principal X-ray line just exceeds the absorption edge of the element to be determined. Curves of spectral purity (in terms of line-to-background ratio) and spectral intensity for various targets using a 2.5 mCi Am²⁴¹ source are presented by Rhodes[5]. The source, a high-intensity γ-emitter, is positioned in a cup formed by the target element. The high-energy γ-radiation from the source excites X-rays characteristic of the target element. These characteristic X-rays, super-imposed over a scattered γ-background, are the sources of excitation for the sample. The characteristic X-rays from the sample pass through the filter and are incident on a NaI(Tl) scintillation crystal. Light photons from the crystal are transmitted by a light guide to the photomultiplier tube.

3. RESOLUTION OF X-RAY SPECTRA

The resolution of X-ray spectra by crystal diffraction, pulse amplitude discrimination, and Ross filters are discussed in this section. The resolution

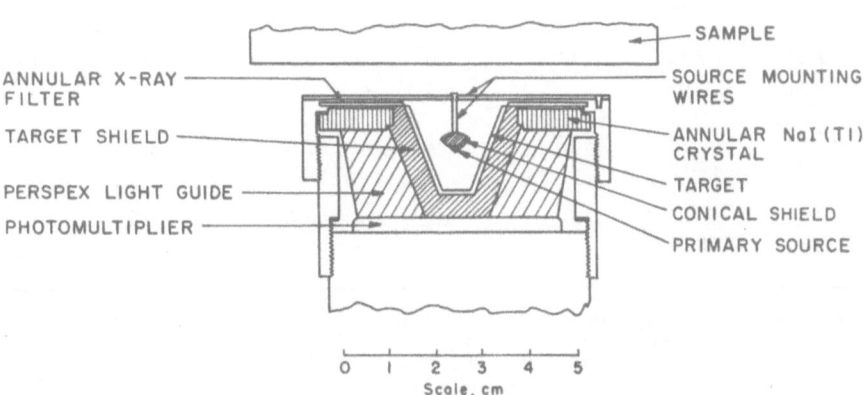

Figure 3. Source–target geometry[5].

of adjacent spectral lines is proportional to the dispersion and inversely proportional to line width. In order to compare the resolution achieved by the various techniques, a figure of merit R equal to the separation of the peaks of characteristic $K\alpha$ lines of adjacent atomic number elements divided by the line width at one half maximum intensity may be used. For Ross filters the line width corresponds to the width of the pass band. The resolving power of conventional flat crystal optics was evaluated for typical collimator–crystal combinations.

Four sets of spectral lines were evaluated covering X-ray energies from Al $K\alpha_1$, 1.49 keV, to Au $K\alpha_1$, 68.8 keV (table II). In the soft X-ray region, < 5 keV, crystal diffraction is 50 to 100 times superior in resolution to either Ross filters or pulse amplitude discrimination using flow-proportional counters. For medium energy X-rays, 5 to 20 keV, crystal diffraction is 2 to 20 times superior in resolution to lithium-drifted silicon; the difference in resolving power between diffraction and lithium-drifted silicon decreases with increasing energy. In the high-energy X-ray range, > 50 keV, the

Table II. Comparison of Spectral Resolution by Diffraction and Energy Dispersion Techniques

Spectral lines	Energy (keV)	$R = D/W\frac{1}{2}$				
		Crystal diffraction	Li-drifted silicon	Proportional counter	Scintillation counter	Ross filter
Al $K\alpha_1$	1.49	43.0[a]		0.3		0.9
Si $K\alpha_1$	1.74					
Ni $K\alpha_1$	7.48	12.0[b]	1.0	0.5	0.2	1.0
Cu $K\alpha_1$	8.05					
Pd $K\alpha_1$	21.18	2.3[b]	1.1	0.5	0.2	1.0
Ag $K\alpha_1$	22.16					
Pt $K\alpha_1$	66.83	1.0[c]	1.3	0.4	0.2	1.0
Au $K\alpha_1$	68.80					

[a] EDdT crystal—coarse collimation, $0.8°2\theta = W\frac{1}{2}$.
[b] LiF crystal—fine collimation, $0.3°2\theta = W\frac{1}{2}$.
[c] LiF crystal—fine collimation, $0.3°2\theta = W\frac{1}{2}$; second-order lines were used.
D = distance between $K\alpha$ peaks for adjacent atomic number elements. $W\frac{1}{2}$ = width of peak at one half maximum intensity. D and W are expressed in 2θ for crystal diffraction and kiloelectron volts for energy dispersion.

lithium-drifted silicon detector results in higher resolution than can be obtained by conventional flat crystal optics. Analyzing crystals with very small d-spacing and high reflectivity are not available so that pulse amplitude discrimination using lithium-drifted silicon and balanced filter techniques are recommended for evaluation in the high-energy X-ray region.

In energy dispersion using pulse amplitude discrimination the experimentally derived peaks are the sum of overlapping spectral lines. In figure 4 the dotted line represents the experimentally measured peak and is the resultant of individual peaks A, B, and C. The problem is to determine at the three positions MA, MB, and MC the relative fraction of the intensities due to the elements A, B, and C. This calculation can be accomplished by either digital or analog techniques; all are based on the solution of simultaneous equations, such as the following:

$$I_{MA} = I_{AA}R_A + I_{BA}R_B + I_{CA}R_C$$
$$I_{MB} = I_{AB}R_A + I_{BB}R_B + I_{CB}R_C \qquad (1)$$
$$I_{MC} = I_{AC}R_A + I_{BC}R_B + I_{CC}R_C$$

where I_{MA}, I_{MB}, and I_{MC} = experimentally determined intensities at the peak positions for A, B, and C on the sample; I_{AA}, I_{BB}, I_{CC} = experimentally determined intensities at peak positions for A, B, C using pure elements; I_{BA}, I_{BC} = experimentally determined intensities of the contribution of elements B and C at the peak position for element A, and R_A, R_B, R_C = amounts of elements A, B, and C in unknown sample. Depending on the complexity of the spectra and the resolving power of the dispersive system, convolution techniques can be used for the determination of major, and under optimum conditions, minor constituents. The reader may consult

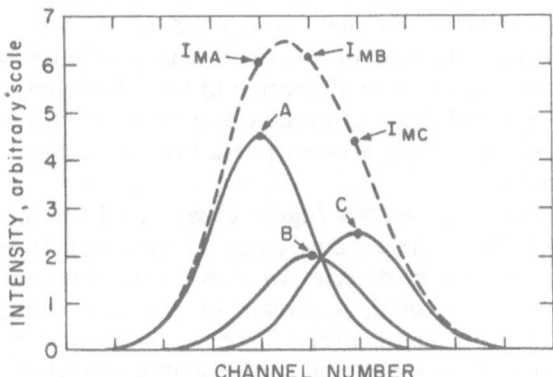

Figure 4. Convolution of X-ray spectra[8].

recent papers on spectral convolution in X-ray analysis for further details[8-11].

Most of the X-ray applications of pulse amplitude discrimination have been limited to the proportional counter because of its superior resolution, as compared to scintillation counters[1]. However, the output pulses from gas-filled detectors are found to be dependent on the incident photon intensity as well as incident photon energy, i.e., the observed pulse amplitude is a function of the counting rate[12,13]. This pulse amplitude shift has been a significant deterrent to more extensive applications of convolution techniques.

3.1. Energy Dispersion using Proportional Counters

The next topic to be considered is detectors, detector–source combinations, and associated instrumentation. By combining a proportional counter–isotopic source assembly with a small multichannel analyzer and strip-chart recorder, Karttunen and his associates developed a very versatile portable X-ray analyzer for the simultaneous determination of several elements[14,15]. This instrument occupies a volume of $\frac{1}{2}$ ft^3 and weighs about 10 lb. Using a Pm^{147} source the range of elements that can be determined includes atomic numbers 19 to 92; greater sensitivity for the low atomic number elements is achieved by the use of an H^3/Zr source.

Tanemura[16,17] used β-ray and β-excited X-ray sources for the excitation of low and medium atomic number elements. These sources are used in conjunction with a thin window proportional counter (figure 5). The scattered β-rays from a Pm^{147} source are prevented from entering the detector by a thin beryllium foil placed over the detector window.

Martin and Blake[18] evaluated various combinations of tritium and metal supports for excitation of low-atomic-number elements. The detector was operated in the flow gas mode using a 0.5-mil-thick beryllium window. With a helium path in place of air, the intensity of the low-energy X-ray lines were significantly increased because of the additional direct excitation by electrons. Spectral lines of carbon and fluorine were measured with their instrumentation. They also include a brief discussion on the safe use of tritium sources.

A continuous gas analyzer (figure 6) was used for the monitoring of SO$_2$, HCl, and Cl$_2$ in the exhaust gases of industrial furnaces [4, p. 147]. To prevent damage to the detector window by the chemically active gases, a double window was used. A stream of gas was continuously flushed through the small opening between these two protective windows.

The use of an α-source in conjunction with a thin window proportional counter (figure 7) is being evaluated for geochemical exploration of the

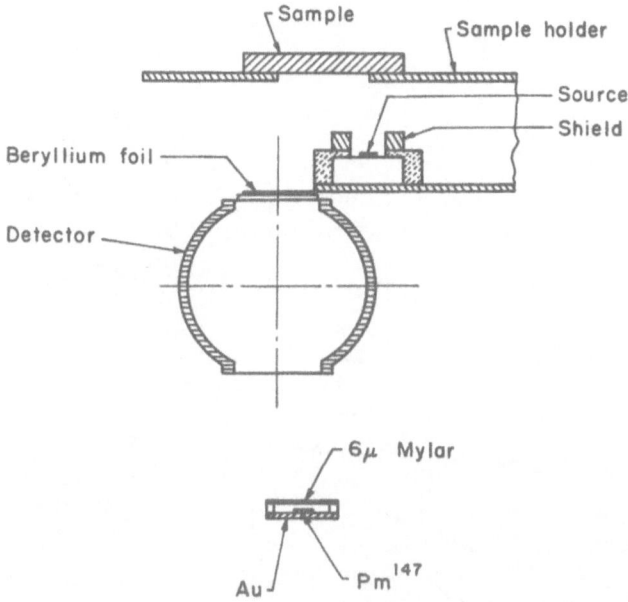

Figure 5. β-source and proportional counter assembly([17]).

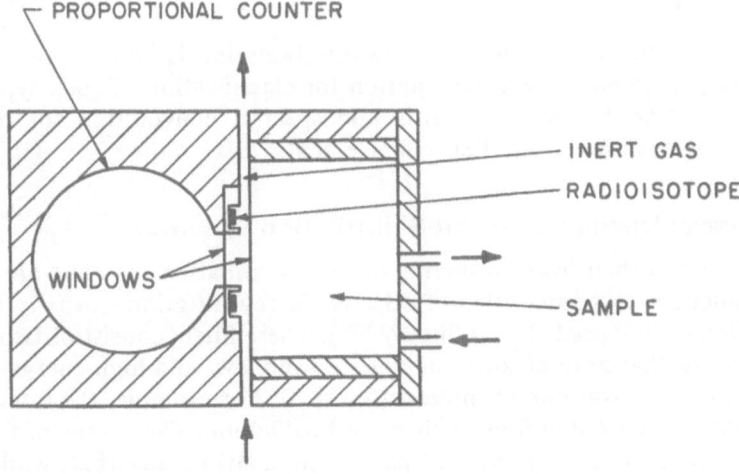

Figure 6. Sample chamber and proportional counter for continuous gas analysis([4]).

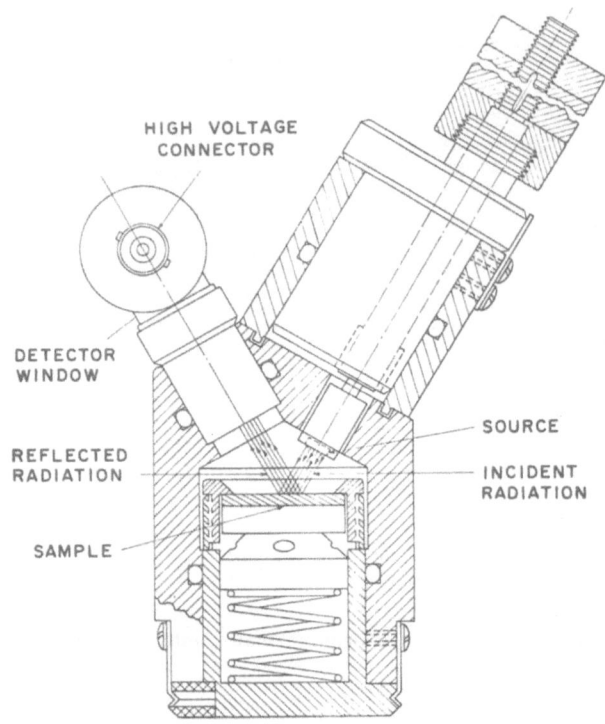

HIGH VOLTAGE
CONNECTOR

DETECTOR
WINDOW

REFLECTED
RADIATION

SAMPLE

SOURCE

INCIDENT
RADIATION

Figure 7. α-source X-ray analyzer for analysis of lunar surface[9].

lunar surface[9]. The problem of data interpretation is being approached in two ways: simple pattern recognition for classification of rock type for sample selection by the astronauts and spectral convolution for determination of concentration of specific elements.

3.2. Energy Dispersion using Scintillation Counters

Because of their lower resolving power, scintillation counters are used in conjunction with some type of selective X-ray filtration—usually some adaptation of balanced (Ross) filters[19,20]. These filters consist of two thin metallic foils that have absorption edges on the low- and high-energy sides of the X-ray spectral line of interest (figure 8). For example, the filters for tin $K\alpha$ radiation are thin foils of silver and palladium. The energy of the tin $K\alpha$ line just exceeds the K edge of palladium, a strong absorber, while the Sn $K\alpha$ X-rays are easily transmitted by the silver foil. The thicknesses of

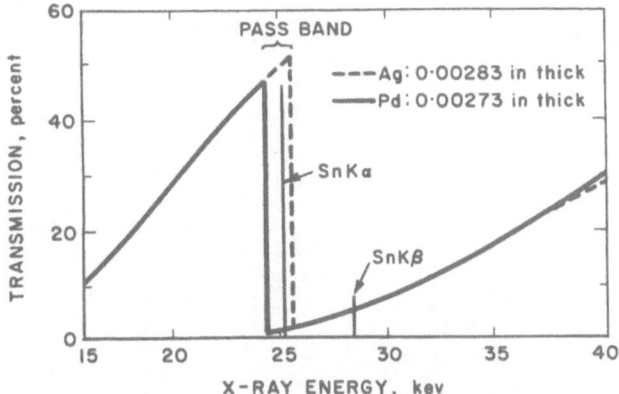

Figure 8. Balanced filters for Sn $K\alpha$ radiation[21].

the two filters are carefully controlled so that their absorption character-istics are similar for all radiation except for the narrow pass band.

The difference in measured intensity, using first the silver and then the palladium filter, is related to the tin content of the sample. Scintillation detectors are adequate for use with Ross filters, since the resolution is controlled by the width of the pass band, rather than the detector resolution. In general, the Ross filter principle is limited to major and, in favorable cases, minor constituents, since it is a difference technique, and the analytical signal is generally a small difference of two large numbers.

Any spectral line whose energy falls within the pass band is a potential source of error. The Ross filter by its nature is a single element analyzer, and each element requires a unique set of filters. The Ross principle is easy to apply, and broad spectrum sources and wide band detectors may be used. Hilger and Watts have developed a commercial model for either laboratory or field use. The laboratory model (figure 9) is similar to the portable model, the latter being more ruggedly constructed and a little lighter (the portable model weighs about 16 lb). We have modified their laboratory model to feed the detector output into a scaler for determination of low concentrations. There are numerous applications to metallurgical and mining problems such as in tin mining and processing[21]. We are presently evaluating the unit for use in the field to determine heavy metals in low-grade ores.

A selective excitation method analogous in principle to Ross filters was proposed by Carr-Brion and Jenkinson[22]. In their modification of a slurry analyzer there are two excitation sources (figure 10). One source emits characteristic X-rays that are of optimum energy for excitation of the element being determined. The other source has a similar spectra, except

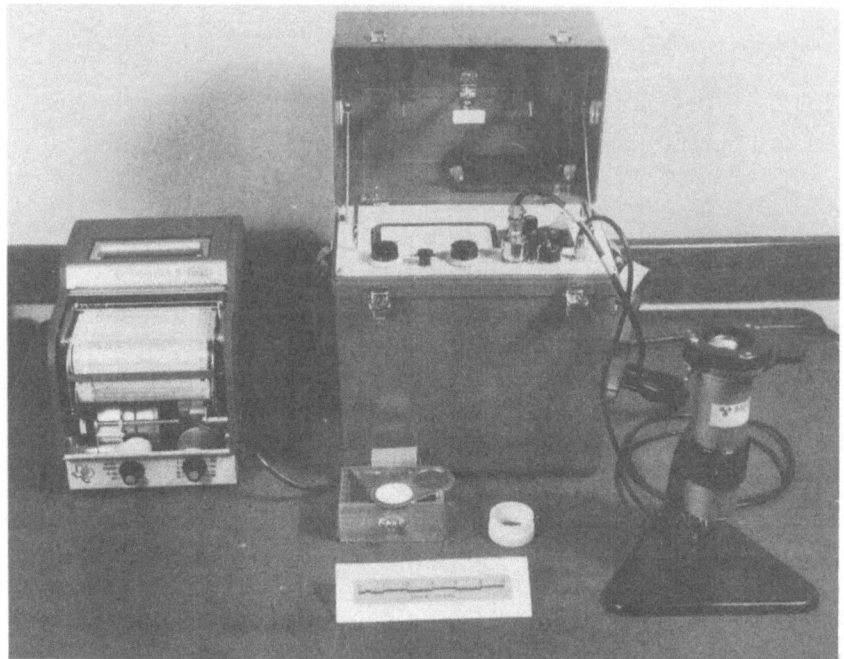

Figure 9. Hilger and Watts laboratory model isotopic analyzer.

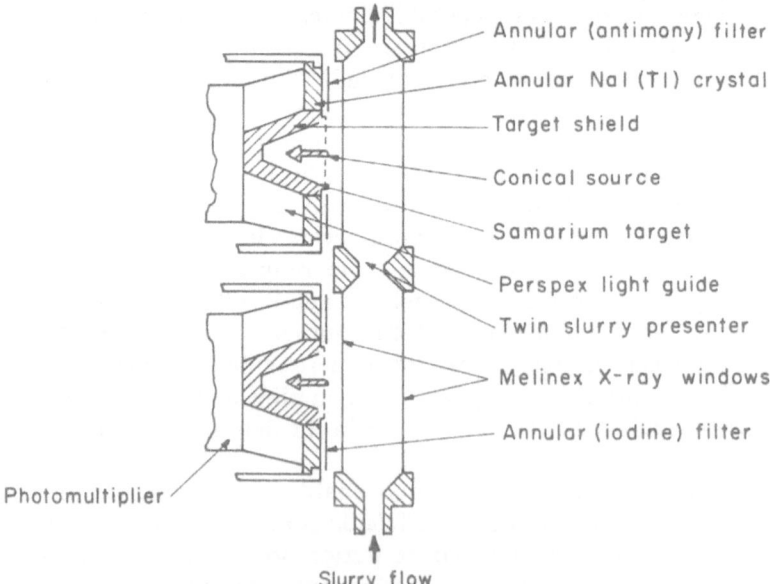

Figure 10. Twin source–target assembly for continuous analysis of slurries[22].

the energy of the characteristic X-rays is close to, but less than, the excitation energy of the element being determined. Thus, the measured radiation from the sample using the first source consists of characteristic plus background, as compared to background only using the second source. The difference in intensity can be directly related to the concentration of the element being determined. For example, samarium $K\alpha$, 40.1 keV, and barium $K\alpha$, 32 keV, targets are used for sources 1 and 2, respectively, for the determination of barium ($K_{edge} = 37.4$ keV). These targets are excited by twin conical Am^{241} sources emitting 60 keV radiation. Zero difference between the backscattered intensities, in the absence of barium in the sample, is achieved by adjusting the distance between the source and secondary target.

Another approach analogous to Ross filters was employed by Brinkerhoff and Forsyth to develop a narrow band X-ray detector[23]. In this instrument all radiation from the sample is first transmitted through the filter and is then incident upon a radiator foil which is an element one atomic number lower than the filter element (figure 11). The pass band is between X-rays whose energy lies between the critical absorption energies of the filter and radiator. When the narrow band detector is combined with a selective source of excitation the instrument is sensitive for a single element such as gold (figure 12).

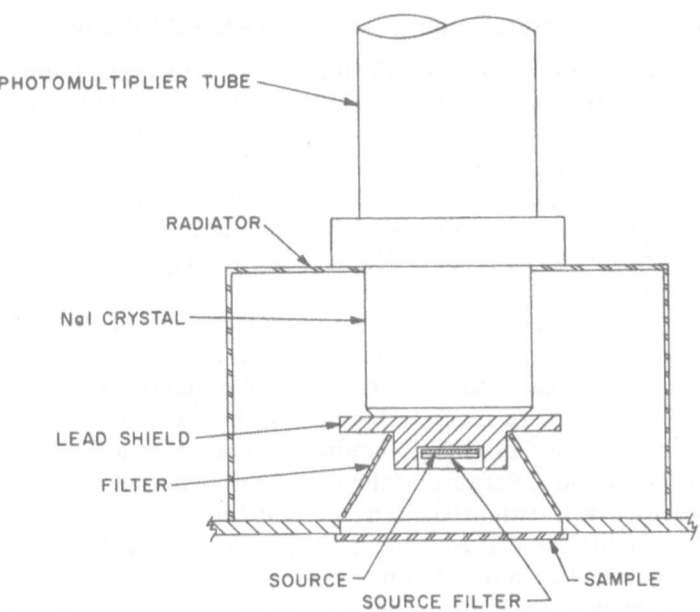

Figure 11. Selective X-ray filter and radiator for energy discrimination[23].

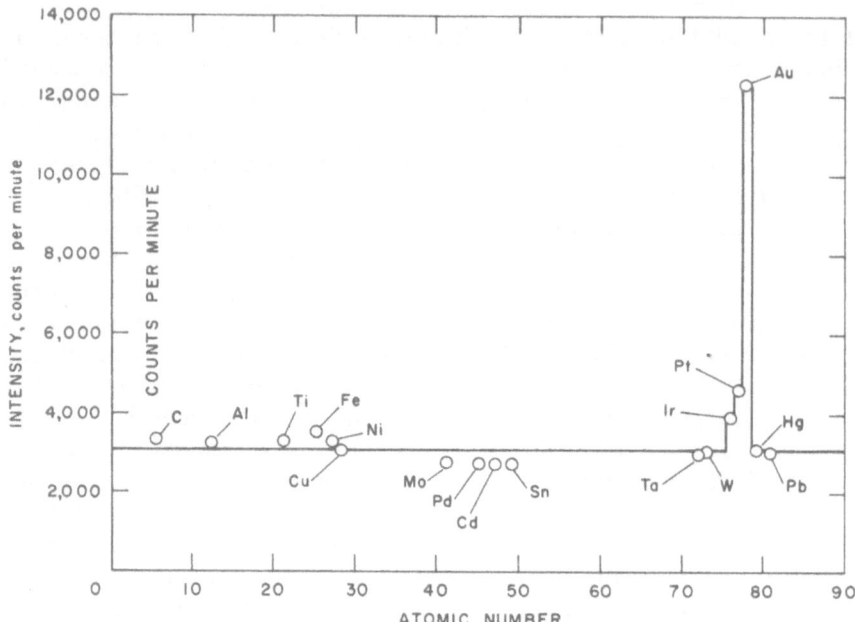

Figure 12. Response of narrow band detector optimized for gold radiation[23].

3.3. Energy Dispersion using Lithium-Drifted Silicon

The advent of the lithium-drifted silicon and germanium detectors represents a major technological breakthrough in γ-ray spectrography[24]. The improvement in resolving power has enabled nuclear researchers to accomplish in one day what previously required, in some instances, most of a year and with less satisfactory results. These solid state detectors are just beginning to be considered for use in the X-ray region, < 100 keV. For Ag $K\alpha$ (22 keV) and Au $K\alpha$ (68 keV), the lithium-drifted silicon detector has pulse widths 2 to 3 times sharper than a proportional detector and 6 to 7 times sharper than the scintillation detector as listed in table II.

Encouraging preliminary results have been obtained using the solid state detector system[25] shown schematically in figure 13. Counting times of several minutes were required because of the low X-ray intensities. The detector–preamplifier assembly requires cryogenic conditions to achieve maximum resolution. Because of the very strict requirements for mounting lithium-drifted germanium crystals, we will limit our applications to the more rugged lithium-drifted silicon. Research is needed to determine if there is any effect of intensity on linearity, pulse width, and pulse shape.

The lithium-drifted silicon detector provides excellent resolution but only moderate detection efficiency in the high-energy X-ray region. The K

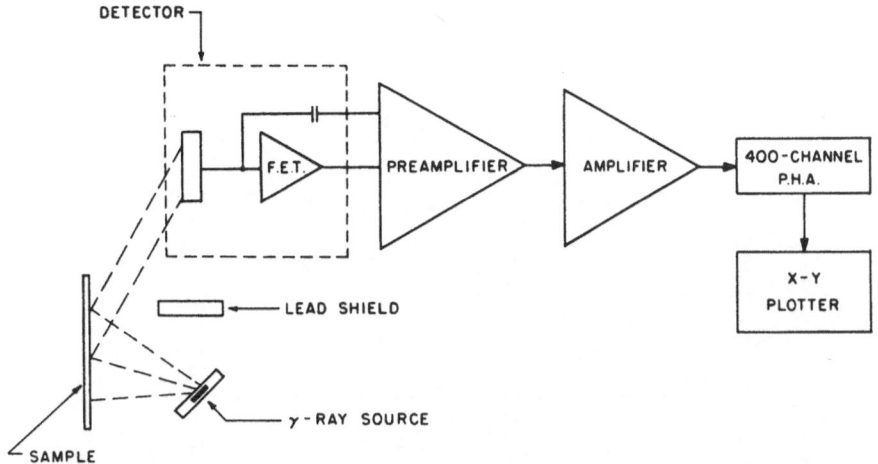

Figure 13. Schematic of energy dispersion analyzer using lithium-drifted silicon or germanium detector([25]).

spectra of high atomic number elements shown in figure 14 were excited by an americium[241] source. Conventional X-ray power supplies are not adequate for efficient excitation of X-rays with energies above 30 to 50 keV. Note in particular in figure 14 the excellent resolution of the $K\alpha_1\alpha_2$ doublets of gold and lead. In addition to the high resolution obtainable, particle size and mineralogical effects can be minimized because the analyst is using very penetrating X-radiation. This increased penetration of high-energy K X-rays has not been exploited for practical analytical problems.

4. APPLICATIONS

The application of energy dispersion instruments are discussed in this section. The principal problem in conventional X-ray spectrography—the matrix effect—is equally important in energy dispersion. The analytical accuracy is directly related to the similarity of standard and unknown or to the validity of the mathematical correction applied to the preliminary results. When the sample is analyzed in the laboratory and the sensitivity is adequate, the matrix effect can be reduced or effectively eliminated by dilution. For *in situ* analysis, or when sample handling is to be minimized, alternative methods are required. Probably the best approach is to use the intensity of backscattered X-rays as a means for matrix correction. Although there have been many successful applications of backscattered X-rays the analyst must first establish the validity of this approach for his particular problem.

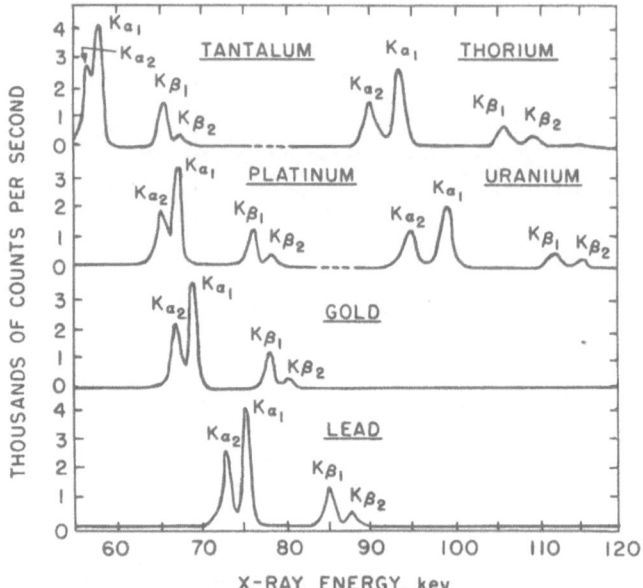

Figure 14. *K* X-ray spectra of high atomic number elements using a lithium-drifted silicon detector[25].

Other analytical problems in conventional X-ray spectrography such as particle-size effect, surface preparation, and chemical effects are also present in energy dispersion but to a different degree. The use of high-energy *K* series X-ray lines will minimize variations in intensity due to surface roughness or particle size. Penetration of the characteristic X-rays will be increased from microns to millimeters. In the soft X-ray region, peak shifts resulting from chemical effects will not be observed because of the much poorer resolution of the energy dispersion method. For example, with a crystal spectrometer the characteristic *K* lines of a low atomic number element will shift off the peak position with a change in chemical valence. In energy dispersion the total *K* intensity is measured rather than intensity at the peak position. Obviously, this poorer resolution is not always advantageous. There are many limitations to applications of energy dispersion imposed by inadequate resolution of spectral lines, particularly when determining low concentrations.

The energy dispersive X-ray systems are being used for many of the research and control analyses now being provided by conventional X-ray spectrographs. Applications to composition and coating thickness are summarized by Rhodes[26] and Clayton and Cameron[4]. The analysis of alloys, cement, coals and coal ash, and industrial plant products have been reviewed[2,4,10,21].

In conclusion, I predict a very rapid increase in the application of energy dispersion X-ray techniques using radioisotope excitation sources. This increase will be a result of the advantages offered by this technique: lower cost, excitation selectivity and stability, portability, and simplicity of operation. These new instruments cost 2 to 5 times less than conventional X-ray spectrographic equipment: therefore, they can be purchased economically for specific problems. There is the opportunity for much greater selectivity in excitation conditions as a wide range of isotopic sources are available at moderate cost. The isotopic sources eliminate the need for voltage and current stabilizers and the electrical and water requirements of the conventional sealed X-ray tube. The battery-operated energy dispersion systems are compact lightweight instruments; thus, the portability can be used to advantage in many applications. Finally, the simplicity of operation should increase the utilization by lower skilled personnel working in metallurgical, chemical, and mining industries and in biological and clinical laboratories.

ACKNOWLEDGMENT

The comments and criticism of Philip G. Burkhalter, particularly those on detector characteristics, are greatly appreciated.

REFERENCES

1. W. Parrish and T. R. Kohler, Use of counter tubes in X-ray analysis, *Rev. Sci. Instr.* **27**, 795–808 (1956).
2. P. S. Baker and M. Gerrard (eds.), Low Energy X- and Gamma Sources and Applications, ORNL-IIC-5 (1965).
3. J. C. Dempsey and P. Polishuk, *Radioisotopes for Aerospace*, Vols. 1 and 2, Plenum Press, New York (1966).
4. International Atomic Energy Agency, Vienna, Austria, *Radioisotope Instruments in Industry and.Geophysics*, Vols. 1 and 2 (1966).
5. J. R. Rhodes, Radioisotope X-ray spectrometry, *The Analyst* **91**, 683–699 (1966).
6. B. Sellers, Generation and practical use of monoenergetic X-rays from alpha emitting isotopes, NYO-3491-1 (1966).
7. L. E. Preuss, A compilation of beta excited X-ray spectra: Part I and Foreword, TID-22361 (1966).
8. L. S. Birks, R. J. Labrie, and J. W. Criss, Energy dispersion for quantitative X-ray spectrochemical analysis, *Anal. Chem.* **38**, 701–707 (1966).
9. J. I. Trombka, I. Adler, R. Schmadebeck, and R. Lamothe, Non-dispersive X-ray emission analysis for lunar surface geochemical exploration, Goddard Space Flight Center, X-641-66-344 (1966).
10. W. J. Campbell, J. D. Brown, and J. W. Thatcher, X-ray absorption and emission, *Anal. Chem.* **38**, 416R–439R (1966).

11. G. Seibel and J. Y. LeTraon, X-ray fluorescence analysis from a radioactive source II, *Revue de Metallurgie* **61**, 342–353 (1964); English translation available as BISI 4054.
12. P. G. Burkhalter, J. D. Brown, and R. L. Myklebust, Pulse amplitude shifts in gas proportional X-ray detectors, *Rev. Sci. Instr.* **37**, 1267–1268 (1966).
13. N. Spielberg, Effect of anode material on intensity dependent shifts in proportional counter pulse height distributions, *Rev. Sci. Instr.* **38**, 291 (1967).
14. J. O. Karttunen, H. B. Evans, D. J. Henderson, P. J. Markovich, and R. L. Niemann, A portable fluorescent X-ray instrument utilizing radioisotope sources, *Anal. Chem.* **36**, 1277–1282 (1964).
15. J. O. Karttunen and D. J. Henderson, An improved portable fluorescent X-ray instrument using radioisotope excitation sources, *Anal. Chem.* **37**, 307–309 (1965).
16. T. Tanemura, X-ray spectroanalysis of light elements using H^3/Zr as exciting source, *Japan J. Appl. Phys.* **3**, 208–214 (1964).
17. T. Tanemura, Performance of an X-ray spectroanalyzer using Pm^{147} beta-ray source, *Japan J. Appl. Phys.* **5**, 51–58 (1966).
18. T. C. Martin and K. R. Blake, A study of fluorescent excitation using isotopic X-ray emission, ORO-627 (1965).
19. P. Kirkpatrick, On the theory and use of Ross filters, *Rev. Sci. Instr.* **10**, 186–191 (1939).
20. P. A. Ross, A new method of spectroscopy for faint X-radiation, *J. Opt. Soc. Am.* **16**, 433–438 (1928).
21. S. H. U. Bowie, A. G. Darnley, and J. R. Rhodes, Portable radioisotope X-ray fluorescence analyser, *Trans. Inst. Mining Met.* (*London*) **74**, 361–378, 557–563, 659–662, 947–948, 978–981 (1965).
22. K. G. Carr-Brion and D. A. Jenkinson, A selective non-dispersive X-ray fluorescence analyser without balanced filters, *Brit. J. Appl. Phys.* **17**, 1103–1104 (1966).
23. J. M. Brinkerhoff and R. Forsyth, Investigation of a new type narrow band X-ray detector for the identification of chemical elements using radioisotope X-ray sources, NYO-3160-1 (1965).
24. J. M. Hollander and I. Pearlman, The semiconductor revolution in nuclear radiation counting, *Science* **154**, 84–94 (1966).
25. H. R. Bowman, E. K. Hyde, S. G. Thompson, and R. C. Jared, Application of high resolution semiconductor detectors in X-ray emission spectrography, *Science* **151**, 562–568 (1966).
26. J. R. Rhodes, Composition and coating thickness testing using radioisotope techniques, AERE-R4457 (1963).

III. QUANTITATIVE ELECTRON MICROPROBE ANALYSIS

Robert E. Ogilvie

Department of Metallurgy, Massachusetts Institute of Technology
Cambridge, Massachusetts

The electron microanalyzer is one of the newest scientific tools applied to the quantitative study of materials. It is capable of supplying the research investigator with experimental data which heretofore was virtually unobtainable.

The applications of the electron microanalyzer have not been limited to any particular area of research. From an examination of the literature, we find that the instrument has been used extensively in the fields of metallurgy, ceramics, mineralogy, and biology—in fact, in all fields in which one finds it necessary to obtain information about the chemistry on a 1-μ scale. Of course, only when the chemistry is examined on such a scale do we discover that the so-called homogeneous materials are not homogeneous after all.

As we shall see, it is not difficult to obtain X-ray data from an area 1 μ in diameter; however, it is another matter to convert this information into a chemical composition. It is necessary to consider what happens to the radiation before it emerges from the surface of the specimen. First, we must consider that the primary intensity is reduced by absorption within the sample; then additional ionizations are produced by the continuous spectrum and by characteristic lines which fall below the absorption edge of the element being measured. Along with these corrections, it is also necessary to take into account the atomic number effect. It is intended to treat these correction procedures in detail and to show that electron microprobe analysis can be quantitative.

1. INTRODUCTION

The electron microanalyzer is one of the newest scientific tools applied to the quantitative study of materials. It is capable of supplying the research investigator with experimental data which heretofore have been virtually unobtainable. As with many instruments in use today, it can be employed with little or no understanding of the principles involved. However, in order to obtain a complete realization of its potentialities, it is necessary to have an elementary knowledge of the physics involved. Without a clear concept of related terms such as electron retardation, fluorescence yield, absorption, and secondary fluorescence, the research investigator not only will be unable to utilize the instrument at its maximum performance, but he will have no understanding of its limitations. Therefore, with only a reference to the excellent presentations by Castaing[1] and others[2,3,4] on the physics of X-ray production, we shall proceed from this point assuming that the reader has a limited knowledge of the subject.

The applications of the electron microprobe analyzer have not been limited to any particular area of research. From an examination of the literature, we find that the instrument has been used extensively in the fields of metallurgy, ceramics, mineralogy, and biology—in fact, in all fields in which one finds it necessary to obtain information about the chemistry on a 1-μ scale. Of course, only when the chemistry is examined on such a scale do we discover that the so-called homogeneous materials are not homogeneous after all.

As we shall see, it is not at all difficult to obtain X-ray data from an area 1 μ in diameter; however, it is another matter to convert this information into a chemical composition. In many cases, it has been found that it is not the mathematical expressions for the absorption, fluorescent, and atomic number corrections that are in serious error, but rather the numerical values of the parameters needed for making the necessary calculations.

1.1. The Electron Microprobe Analyzer

The analysis of X-ray spectra excited by electrons is an outgrowth from the early work of Moseley[5]. Utilizing the technological advances in electron optics, which were made by Ruska, Haine, Zworkin, Hillier, and others, Castaing and Guinier[6] were able to construct an electron optical column that produced a finely focused electron beam 1 μ in diameter on the surface of the specimen. This electron optical column, along with a suitable specimen chamber, was then integrated with an X-ray spectrometer and a light-optical system. These features constitute the most important aspects of the electron microprobe analyzer.

The electron optical column, illustrated in figure 1, produces a demagnified image of the electron source (the crossover of the electron gun) on the surface of a well-polished specimen. The source size is usually about 50 to 100 μ. The electron optical column reduces the beam size by a factor of about 100 to produce a 0.5 to 1.0 μ image of the source on the surface of the specimen. These incident electrons produce an X-ray spectrum that contains characteristic lines from all the elements present in the irradiated specimen volume. With the aid of the X-ray spectrometers, the spectra are dispersed and analyzed for their wavelength and intensity.

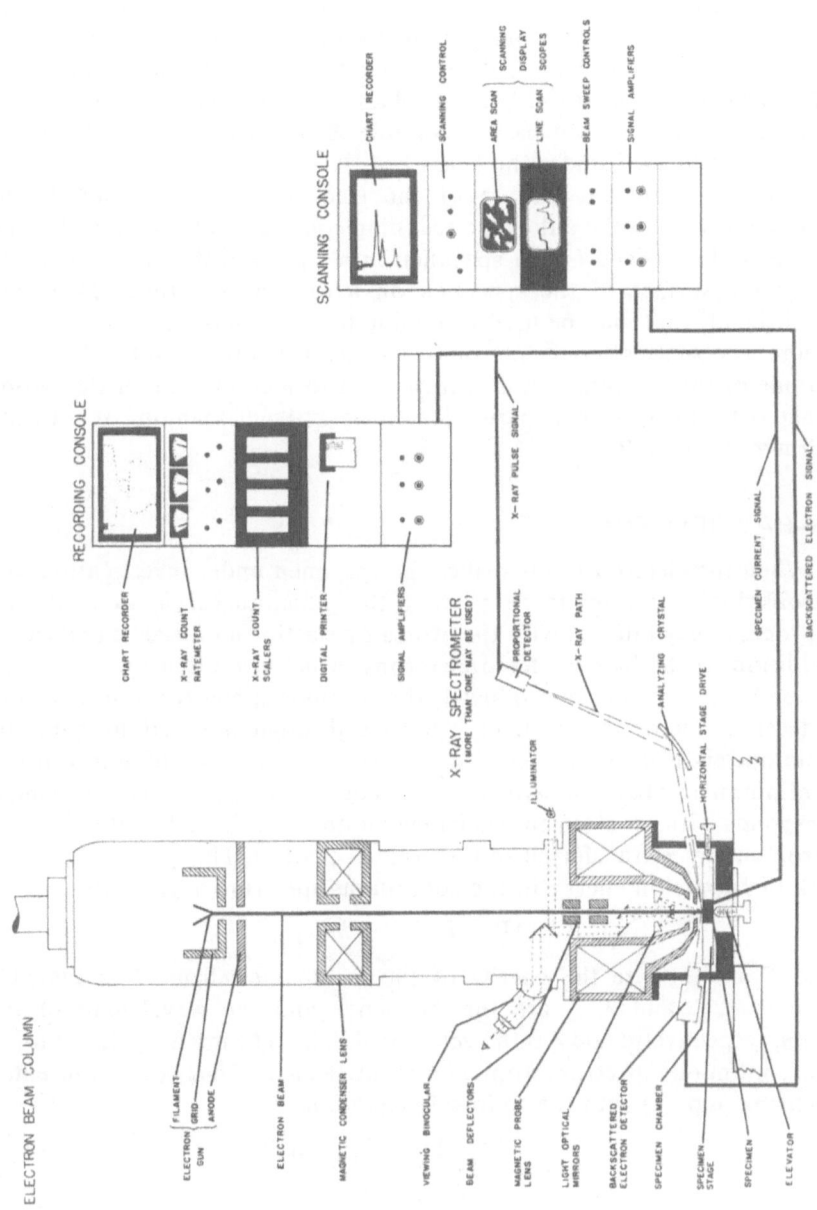

Figure 1. Schematic drawing of the electron optical column and associated consoles.

It is vital that the operater be able to observe the irradiated area with a light optical system. This will enable him to analyze with confidence areas not much larger than the beam. The light optics are also necessary to relocate areas that were previously studied with a conventional metallurgical or petrographic microscope and, in particular, areas of interest indicated with fiducial marks made with a microhardness indenter. This means that the specimen stage should have a suitable X–Y drive so that these small areas may be located under the beam.

Another important feature of the electron optical column is the scanning coils. With the aid of the scanning coils, which sweep the electron beam over the surface of the specimen in a square raster, one is able to observe the surface of the specimen on a cathode ray tube. There are several signals that may be used to modulate the brightness of the cathode ray tube: (1) backscattered electrons, (2) specimen current, or (3) the signal from one of the X-ray detectors which is set to measure one of the strong characteristic lines from a particular element. Such scanning modes are illustrated in figure 2.

2. X-RAY SPECTRA

When the electron beam strikes the specimen under investigation, the electrons diffuse in a manner similar to that illustrated in figure 3; during this process, they interact with the atoms by elastic and inelastic collisions. In addition to the backscattered electrons, which are discussed later, two types of X-ray spectra are produced; the continuous spectrum results from the deceleration of the incident electrons through interactions with the atomic nuclei. Because the electrons have a wide range of interactions, a continuum is observed with a sharp cut-off at λ_{min}. This wavelength corresponds with an electron–nuclei encounter in which all of the incident electron energy is transferred to the emitted photon. This particular wavelength of maximum energy in the continuous spectrum is given by

$$E = eV = h\nu_{max} = hc/\lambda_{min} \tag{1}$$

where E and eV are the energy of the incident electron; h is Planck's constant; ν_{max} and λ_{min} are the frequency and the wavelength of the photon, respectively; and c is the velocity of light. The intensity distribution of the continuous spectrum from a target of element Z has been represented by Kulenkampff[7] with the following equation:

$$I_\lambda = \frac{MZ}{\lambda^2}\left[\frac{1}{\lambda_{min}} - \frac{1}{\lambda}\right] + \frac{NZ^2}{\lambda^2} \tag{2}$$

where M and N are constants. As the energy of the incident electrons is

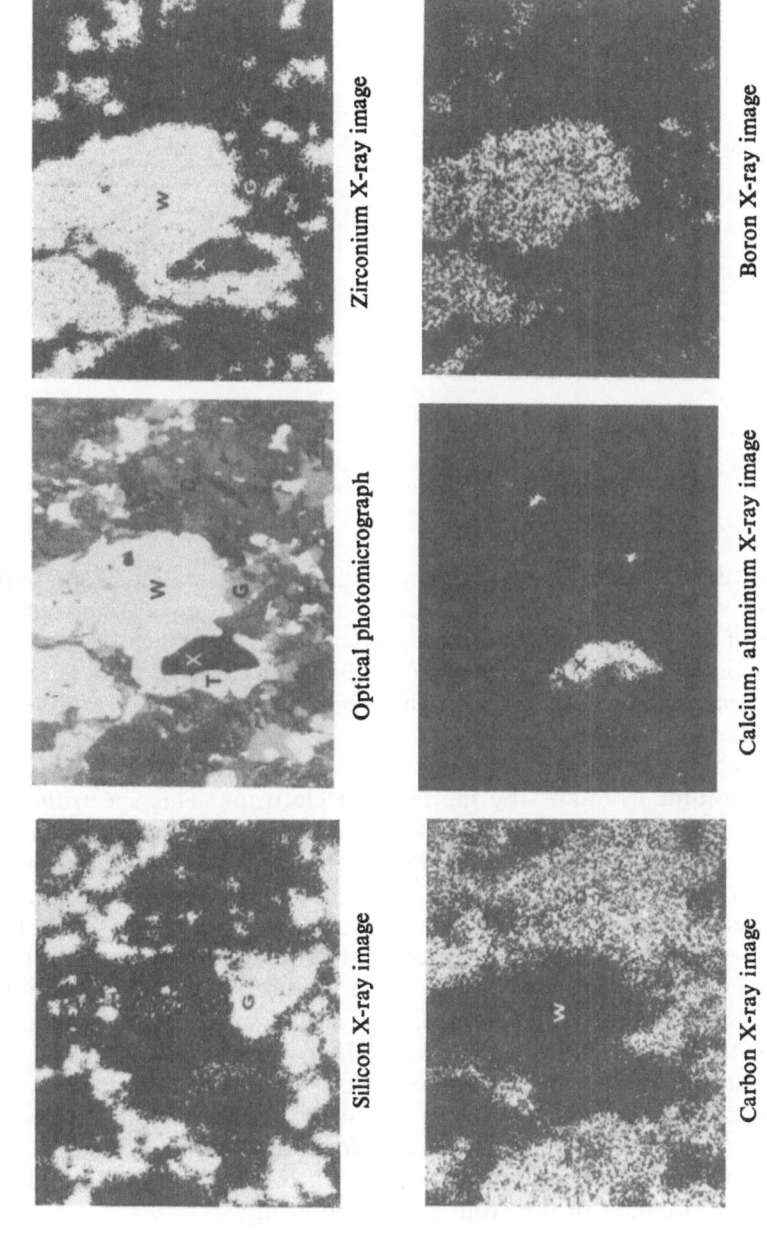

Microconstituent Phases in JTA Graphite   48C,35Zr,8B,9Si

Zirconium X-ray image

Optical photomicrograph

Silicon X-ray image

Boron X-ray image

Calcium, aluminum X-ray image

Carbon X-ray image

Figure 2. Scanning display photographs of the microconstituent phases in JTA graphite.

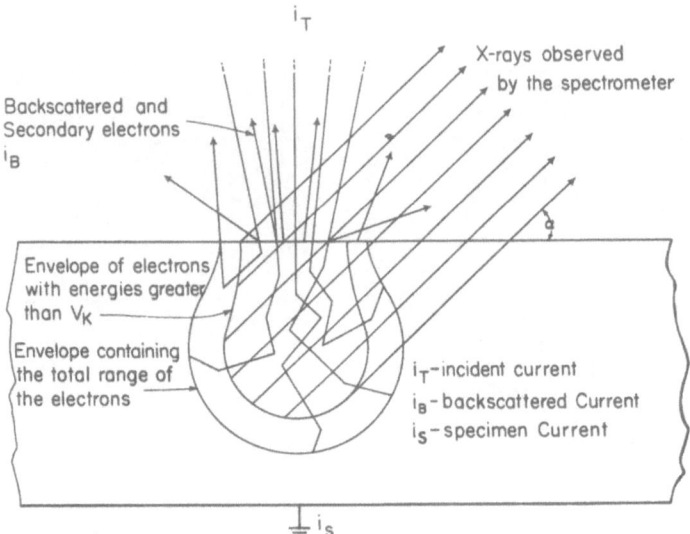

Figure 3. Schematic representation of the diffusion of electrons into the specimen.

increased, the intensity of all wavelengths increases and λ_{min} shifts to shorter wavelengths. Neglecting the second part of eq. (2) and using the result of eq. (1), we see that for a particular wavelength λ the intensity is proportional to the energy (eV) of the incident electron. This latter point is important when considering the signal-to-background ratio for the characteristic lines.

A characteristic spectrum is produced for each kind of atom present in the volume irradiated by the incident electrons. This spectrum is best explained by the energy diagram of the atom, which is illustrated in figure 4. The K excitation state corresponds to the work necessary to eject a K electron from the atom. Similarly, the L, M, and N states correspond with the ejection of the outer-bound electrons. An atom in any one of these excited states will return to the ground state in about a nano-nano second. The return to the ground state is made by discrete energy transitions, as indicated. Each transition corresponds to an emission of a particular photon. This is the result of an outer electron falling into the inner shell which has lost an electron. As these energy states are extremely sharp, the energy of the photon is extremely sharp. The wavelength λ of a photon corresponding to a particular type of transition, e.g., from the K to the L_{III} state (which corresponds to the $K\alpha_1$ photon), was shown by Moseley[5] to be related to the atomic number of the atom by the following:

$$\lambda = P(Z - \sigma)^{-2} \tag{3}$$

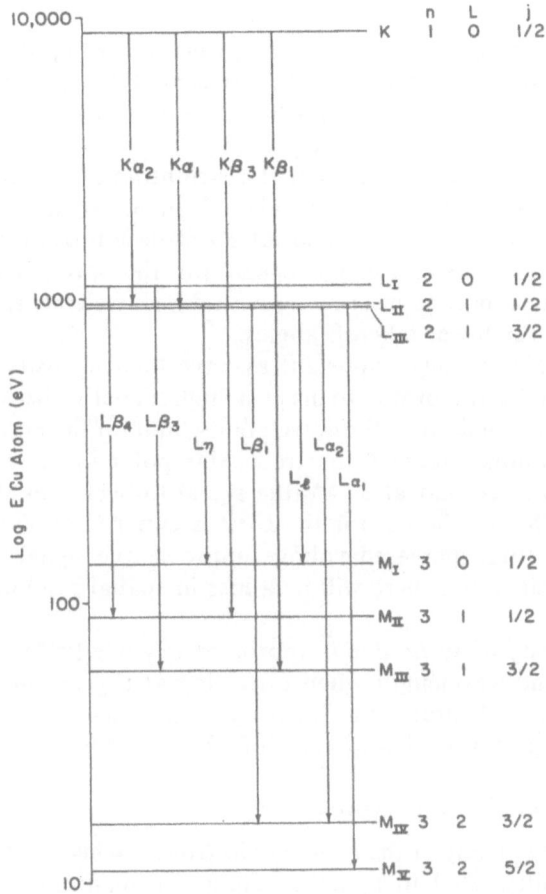

Figure 4. Energy diagram of the copper atom. Only the important transitions are indicated.

where P is a constant for the particular type of transition and σ is a screening constant and equal to one for the K X-ray series.

In electron microanalysis, the $K\alpha$ spectrum is used for the elements from boron (5) through bromine (35), while the $L\alpha$ spectrum is used from krypton (36) through uranium (92). However, improved signal-to-background ratio may be realized by using the $M\alpha$ spectrum for the elements from gold (79) through uranium (92). It has been observed by Brown and Ogilvie[9] that the intensity of a $K\alpha$ spectrum line is given by the following

$$I = Qi(V_0 - V_K)^{1.67} \tag{4}$$

where Q is a constant, i is the incident electron current; V_0 is the energy of the incident electrons; and V_K is the energy necessary to put the atom in the K excitation state. From eqs. (4) and (2) we see that the signal-to-background ratio for a K spectrum line is proportional to V^n, where n is approximately 0.7 for incident electron energies equal to $3V_K$.

It is interesting to note the spatial distribution of a characteristic line, i.e., the variation of intensity with take-off angle. Figure 5 illustrates the angular distribution for Cu $K\alpha$ and Al $K\alpha$ radiation from pure standards at 30 kV. As the absorption coefficient for the wavelength of interest increases, for instance, as in alloying with a heavy element, the intensities are reduced at the lower take-off angles.

While it is extremely important to have high intensity characteristic lines, it is equally important to have a high signal-to-background ratio, since these factors determine the detectability limit for the particular element under investigation. Figure 6 illustrates this point in the case of molybdenum; here we see that at 30 kV the signal-to-background ratio for the $K\alpha$ is at best 15 : 1, whereas for the $L\alpha$ it is better than 300 : 1. It should be pointed out that increased voltage improves the signal-to-background ratio in both cases, but there will be a loss in spatial resolution within the specimen.

The intensity of an $L\alpha$ line is approximately one half that of a $K\alpha$ line having the same wavelength when operating at a given incident electron energy. Figure 7 illustrates the intensity of a series of K and L lines at different voltages for a take-off angle of 52.5°.

2.1. Backscattered Electrons

We find that many of the incident electrons are backscattered from the specimen, as illustrated in figure 3. These are made up of two types—

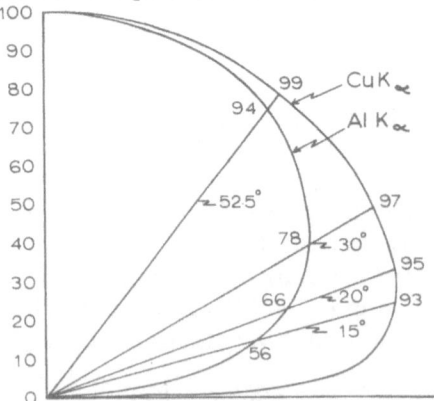

Figure 5. Spatial distribution of copper and aluminum $K\alpha$ radiation at 30 kV.

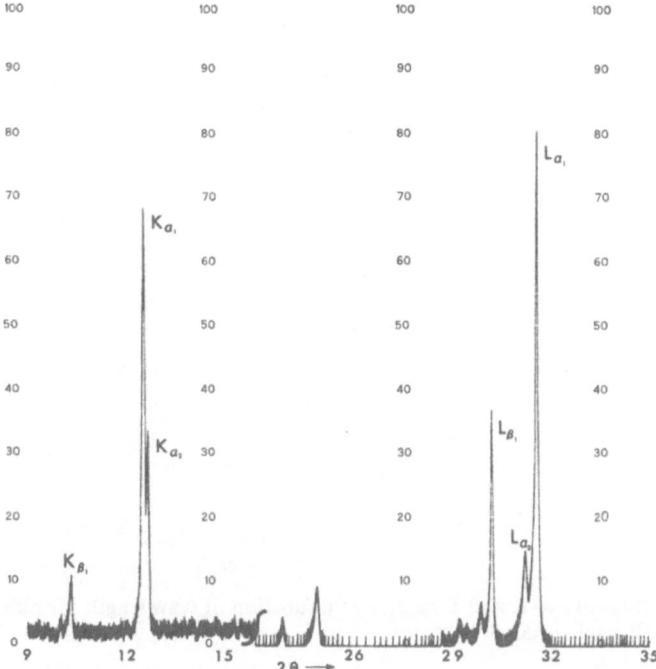

Figure 6. The K and L spectra from molybdenum at 30 kV and 15.5° take-off angle using a mica crystal and a flow proportional counter (for the K series, 400 c/sec full-scale and for the L series, 3200 c/sec full-scale).

primary and secondary electrons. The secondary electrons are those with energies less than 50 eV. These low-energy electrons, which will not be considered here, have proved very useful in the study of potential gradients in microelectronic circuits[8] and are generally used in scanning electron microscopy. .

The energy distribution of the primary backscattered electrons is not unlike that of the continuous X-ray spectrum in that there is a sharp cut-off at the energy of the incident beam. The energy distributions as calculated by Brown and Ogilvie[9] for Al, Cu, and Au are in good agreement with the experimental measurements shown in figure 8.

The backscatter coefficient is $(i_b - i_s)/i_b$, where i_b is the incident beam current and i_s is the specimen current. Sternglass[10] and others have shown that its values depend on the atomic number of the irradiated element. The backscattered fraction for aluminum is 0.12, and for uranium it is about 0.55. The dependence on atomic number suggests the possibility of

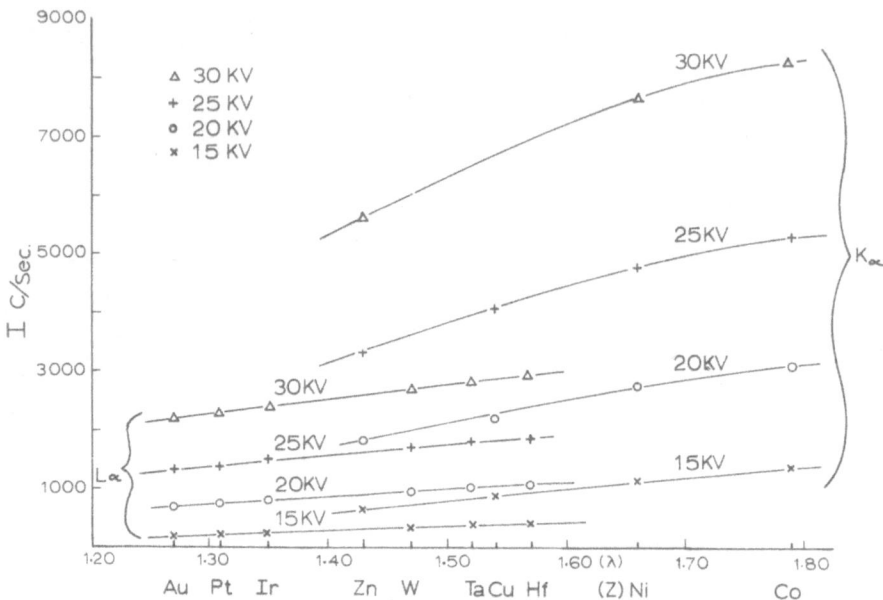

Figure 7. Intensity of K and L spectra as a function of wavelength for different voltages at a take-off angle of 52.5°.

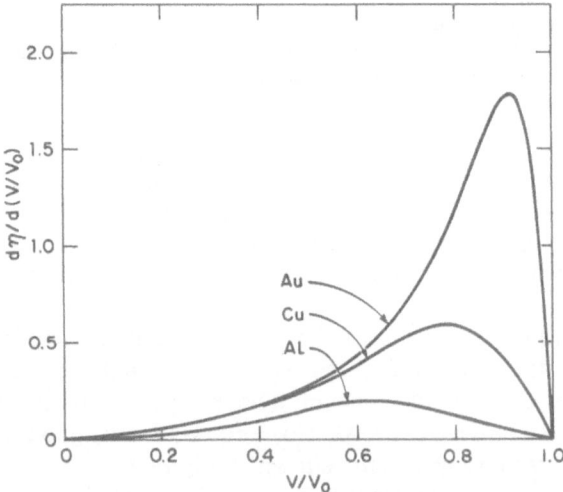

Figure 8. Energy distribution of the primary back-scattered electrons from Al, Cu, and Au. The area under the curves represents the fraction backscattered. For Al, Cu, and Au, this is 0.12, 0.29, and 0.46, respectively.

using the primary backscattered electrons as a means of chemical analysis. Ogilvie and Lewis[11] presented results obtained from an Al–Zn diffusion couple which demonstrated the use of a high-resolution electron microanalyzer. The fact that the primary backscattered electrons come from an area smaller than that for the X-ray signal is observed in backscattered electron scanning images.

For the analysis of a binary alloy, Castaing[1] has suggested that the backscatter coefficient follows the linear relation

$$\eta_{A+B} = C_A \eta_A + (1 - C_A)\eta_B \tag{5}$$

where C_A is the weight fraction of component A. Wittry[12] has suggested the use of the following expression:

$$\eta_{A+B} = 1 - \frac{(1 - \eta_A)(1 - \eta_B)}{C_A(1 - \eta_B) + C_B(1 - \eta_A)} \tag{6}$$

Neither of these equations has met with much success in quantitative analysis. However, it has been shown by Ogilvie and Lewis[11] and by Ziebold[13] that for binary systems a calibration curve may be made from a series of standard alloys. The calibration curve used by Ziebold consists of a normalized function (i_N) which varies from 0 to 1 for C_A equal to 0 and 1, respectively:

$$i_N = \frac{\eta_{A+B} - \eta_B}{\eta_A - \eta_B} = \frac{i_{A+B} - i_B}{i_A - i_B} \tag{7}$$

where i_A, i_B, and i_{A+B} are the measured specimen currents. Calibration curves of the type i_N vs. C_A will fit the analytical expression

$$\frac{1 - i_N}{i_N} = b_{AB}\frac{1 - C_A}{C_A} \tag{8}$$

where b_{AB} is a constant for the particular A-B binary. It is unfortunate that an exact solution of the chemistry can be made only for a binary or pseudobinary system. We see that equation (8) leads to the following equation for the backscatter coefficient:

$$\eta_{A+B} = \frac{C_A\eta_A + b_{AB}(1 - C_A)\eta_B}{C_A + b_{AB}(1 - C_A)} \tag{9}$$

which reduces to eq. (5) when $b_{AB} = 1$.

The measurement of backscattered electrons has not been used to any great extent as an analytical technique. However, the effect has proved useful in establishing whether a second phase has an average atomic number lower or greater than the matrix.

Application of the backscattered electrons with a high resolution (0.1 μ) electron microanalyzer for multiphase diffusion couples has yet to be demonstrated.

2.2. Intensity of Characteristic X-Radiation

We have seen that the wavelength of the characteristic line establishes the particular element and that the concentration of this element must be deduced from the intensity of its characteristic line. The intensity of characteristic X-radiation has been treated by Worthington and Tomlin[14], Archard[15], Green and Cosslett[16], and by Brown and Ogilvie[17].

The number of K quanta (ν_K) per incident electron which is radiated into the solid angle $\delta\omega$ at a take-off angle α to be measured by the X-ray spectrometer, is given by the following:

$$\nu_K = (\delta\omega/4)N_K$$

$$N_K = \kappa \int_0^{x_K} \exp(-\mu_K x \csc \alpha) \, \partial n/\partial x \, dx \qquad (10)$$

where μ_K is the linear absorption coefficient for the K photons; $\partial n/\partial x$ is the number of K shell ionizations per unit electron path and is equal to NQ; N is the number of atoms per unit volume, and Q is the cross section for K shell ionizations; x_K is that path length where the energy of the electron drops below the energy for K shell ionization; and κ is a correction factor containing (1) a correction for secondary K shell ionization by the continuous X-radiation, (2) the fluorescent yield factor W, (3) correction for the loss of ionization due to backscattered electrons R, and (4) a correction of the absorption factor due to the fact that the electrons do not travel a straight line path.

Rearranging eq. (10) leads to

$$N_K = WR\left[\exp(-\mu_K \epsilon\bar{x}\csc\alpha)\right]\int_0^{x_K} \partial n/\partial x \, dx + N_{K_f} \qquad (11)$$

where $\epsilon\bar{x}$ is defined as the average depth of K photon production, and N_{K_f} is the contribution due to secondary ionization such as that produced by a strong characteristic line from a second element and in part by the continuous spectrum. The term in the square brackets is Castaing's[1] absorption correction $f(\chi)$. Assuming an electron retardation law similar to that of the Thompson–Widdington[18] and Webster[19] laws,

$$dE/dx = -CE^i \qquad (12)$$

equation (11) takes the following form:

$$N_K = WRf(\chi)N\underset{\sim}{Q}x_K + N_{K_f} \tag{13}$$

where $\underset{\sim}{Q}$ is an average cross section for K ionization.

If the ratio of the intensity from the alloy $I_{A+B+\cdots i}$ to the intensity of pure A is defined as K_A, it follows from eq. (13) that

$$K_A = \frac{WRf(\chi)N\underset{\sim}{Q}x_K + N_{K_f}}{WR^0f^0(\chi)N^0\underset{\sim}{Q}x_K^0 + N_{K_f}^0} \tag{14}$$

where the superscripts are for the pure element A.

Equation (14) takes into account all factors which need to be considered for an exact solution of the intensity ratio. However, because of the difficulty in evaluating the loss of ionizations due to backscattered electrons (R), the average ionization ($\underset{\sim}{Q}$), it has been more useful to reduce eq. (14) to the following form:

$$K_A = \left[\frac{f(\chi)}{f^0(\chi)}\right]\left[1 + \frac{I_{sf}}{I_{pf}}\right]\left[\frac{\alpha_A}{\sum\limits_i \alpha_i C_i}\right]C_A \tag{15}$$

where the first bracketed term is Castaing's[1] absorption correction: the second bracketed term is Castaing's[1] secondary fluorescent term: and the third bracketed term is the atomic number correction of Poole and Thomas[20].

Philibert has put Castaing's[1] absorption correction into the following analytical form:

$$f(\chi) = \frac{1 + h}{(1 + \chi/\sigma)[1 + h(1 + \chi/\sigma)]} \tag{16}$$

where $\chi = (\mu/\rho)_{A+B\cdots i}$ csc α, $h = 1.2\Sigma\alpha_i A_i/(\Sigma\alpha_i Z_i)^2$ (α_i being the atomic concentration, A the atomic weight, and Z the atomic number) and σ is the cross section for absorption of the electrons, which may be evaluated from the following:

$$\sigma = 1820(30/V)^2 \tag{17}$$

or

$$\sigma = 2.4 \cdot 10^5/(V^{1.5} - V_K^{1.5}) \tag{18}$$

The latter equation, suggested by Duncumb[21], justifies Castaing's aluminum curve made with a copper tracer and Green's[22] aluminum curve made with an aluminum tracer. Equation (18) is certainly the better choice. The important feature of Philibert's equation is that it is applicable over the voltage range normally employed in electron microanalysis and also covers the range of atomic numbers of the periodic table. Adler and

Goldstein[23] have made available a tabulated form of eq. (16) covering a wide range of voltages. These tables made calculation of the absorption correction extremely simple.

Brown and Ogilvie[9] have shown that it is possible to calculate the $f(\chi)$ curve for any element having any particular tracer. An example of these $f(\chi)$ curves for the copper aluminum system is illustrated in figure 9. When measuring copper in Cu–Al alloys, it is necessary to interpolate between the limiting curves of Cu(Cu) and Al(Cu); when measuring aluminum, one would interpolate between the limiting curves Cu(Al) and Al(Al). At present we can only assume a linear interpolation between the two curves. It should be noted that we do not have adequate information to use the desired limiting curves for all possible systems.

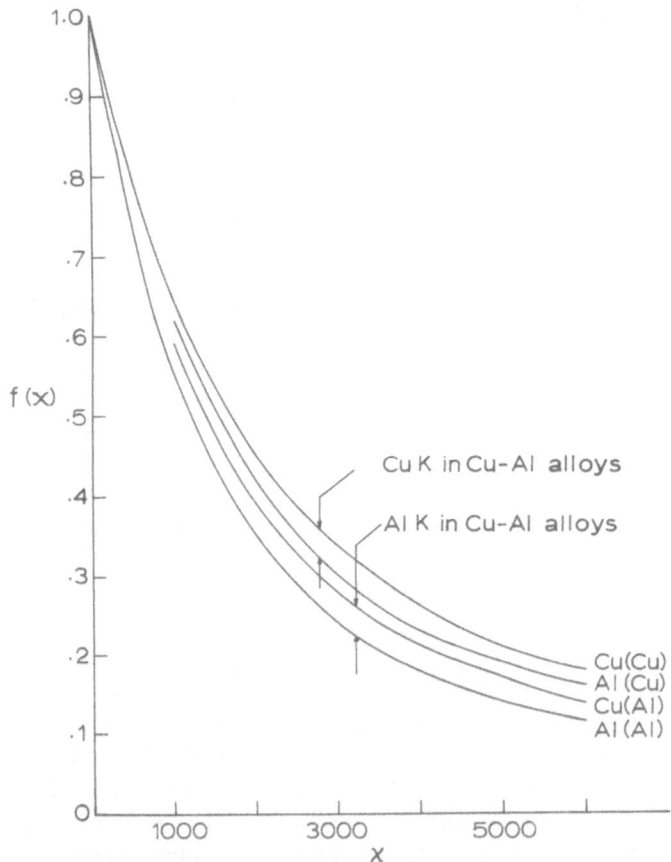

Figure 9. Absorption correction curves for Cu–Al alloys.

The fluorescent correction, the more complicated term in eq. (15), has been derived by Castaing([1]) and is given by the following:

$$\frac{I_{sf}}{I_{pf}} = \frac{W_K^B}{2} \frac{r_A - 1}{r_A} \frac{\lambda_B}{\lambda_A} \frac{A_A}{A_B} \left[\frac{C_B}{C_A + C_B(\mu/\rho)_B^{BK\alpha}/(\mu/\rho)_A^{BK\alpha}} \right]$$

$$\times \left[\frac{\ln(1+u)}{u} + \frac{\ln(1+v)}{v} \right] \qquad (19)$$

where W_K^B is the fluorescent yield factor of element B; r_A is the ratio of the absorption coefficients at the K edge of A; λ_A and λ_B are the wavelengths of the K edges of A and B respectively: A_A and A_B are the atomic weights of A and B; and u and v are given by

$$u = \frac{(\mu/\rho)_{A+B}^{AK\alpha}}{(\mu/\rho)_{A+B}^{BK\alpha}} \csc \alpha$$

$$v = \frac{\sigma}{(\mu/\rho)_{A+B}^{BK\alpha}} \qquad (20)$$

In deriving eq. (19), Castaing assumed that the intensity of the directly excited K radiation of elements A and B could be obtained from the Rosseland function. This assumption has proved to be quite accurate for K lines fluorescing K lines; however, when L lines fluoresce K or L spectra, it is necessary to have a relationship between the intensity of the fluorescing radiation and the primary intensity for these different cases. Since only lines near absorption edges contribute to the fluorescent effect, which is usually small when L's fluoresce K's, it may be possible to show that there is a constant factor between the primary and fluorescent intensity. This is certainly an area for more theoretical and experimental work. However, a possible solution to the dilemma would be the following:

$$\frac{I_{sf}}{I_{pf}} \doteq \frac{W_i^A}{2} \frac{r_A - 1}{r_A} \frac{\lambda_B}{\lambda_A} p^A C_A K_{ji} \frac{f(\chi)}{f^0(\chi)} [XY] \qquad (21)$$

where

$$X = \frac{C_B}{C_A + C_B(\mu/\rho)_B^{BK\alpha}/(\mu/\rho)_A^{BK\alpha}}$$

$$Y = \frac{\ln(1+u)}{u} + \frac{\ln(1+v)}{v}$$

Equation (21) is similar to eq. (19), and the difference is the fluorescent yield factor (W_i^A) is for the element being fluoresced, which may be a K, L, or M line. P^A is the power in the line i, and K_{ji} represents the ratio of intensity of the j line fluorescing i; the other terms have been defined previously.

The latter term of eq. (15), the atomic number correction, takes into account the backscatter effect as a function of atomic number and the retardation of electrons within the sample. The difficulty of evaluating the α parameters from theory or experiment has led to much confusion: however, the treatment of Poole and Thomas[20] is the best available at present.

An expression similar to eq. (10) may be derived where the integration is carried out over the energy of the electron, i.e.,

$$N_\kappa = \text{No } C_A/A \int_0^{x_r} \rho Q_\kappa \, dx$$

$$N_\kappa = \text{No } C_A/A \int_{E_0}^{E_\kappa} \frac{\rho Q_\kappa}{dE/dx} \, dE \tag{22}$$

where No is Avogadro's number; A is the atomic weight; ρ is the density; dE/dx is the loss of energy per unit path length; and Q_κ is the ionization cross section. Then from eq. (22) we may write an expression for the *generated* ratio of X-ray intensities:

$$\frac{I'_{AB}}{I'_A} = C_A \frac{R_{\text{alloy}}}{R_{\text{pure}}} \frac{\displaystyle\int_{E_\kappa}^{E_0} Q_\kappa/S_{\text{alloy}} \, dE}{\displaystyle\int_{E_\kappa}^{E_0} Q_\kappa/S_{\text{pure}} \, dE} \tag{23}$$

where R takes into account the loss of ionizations due to backscatter and S is the stopping power $[1/\rho(dE/dx)]$. This latter term is determined by the mass/unit area.

Castaing[1] proposed in his second approximation that the ratio of the *generated* intensities is related to the mass concentrations by the following:

$$\frac{I'_{AB}}{I'_A} = C_A \alpha_A / \sum_i C_i \alpha_i \tag{24}$$

where α_i depends on the atomic number of the element i.

If we use Webster's law,

$$dE/dx = (\text{const}) Z_i \, \rho/A_i \tag{25}$$

eq. (23) may be solved to yield the following results:

$$\alpha_A = (1/R)_A (Z/A)_A \quad \text{and} \quad (1/R)_{\text{alloy}}(Z/A)_{\text{alloy}} \tag{26}$$

Since the relative retardation and backscattering loss do not have the same dependence on atomic number, eq. (24) will not yield the correct results.

Thomas has defined the α's in eq. (15) as follows:

$$\alpha_{\text{alloy}} = S_{\text{alloy}}/R_{\text{alloy}} = \sum_i (S_i/R_i)C_i \qquad (27)$$

Graphs of R and S vs. atomic number have been prepared by Thomas[24], and this is the best approach that we have at present in spite of its short-comings.

3. EMPIRICAL APPROACH

In the previous section it was shown that quantitative analysis can be performed in multicomponent specimens. The only essentials are pure standards and intensities above the background of characteristic lines from each element in the alloy.

Due to inadequate data on absorption coefficients, fluorescent yields, jump ratios, and approximations used in the derivation of the analytical expressions, e.g. (15) is not applicable to many systems. Because of this shortcoming, Ziebold and Ogilvie[25] presented an empirical approach which requires the use of the pure elements and $n - 1$ standards for an n-component system. It has been found experimentally that the calibration curves of binary systems are well defined by the following equation:

$$\frac{1 - K_A}{K_A} = a_{AB}\frac{1 - C_A}{C_A} \qquad (28)$$

This single conversion parameter has been correlated with many binary systems and may be expressed as

$$a_{AB} = 0.95\left[\frac{\sigma + \chi_B^A}{\sigma + \chi_A^A}\right]\left[\frac{Z_A}{Z_B}\right]^{0.3}\left[\frac{1}{1 + 0.07\kappa}\right] \qquad (29)$$

where κ is a simplified expression for secondary fluorescence

$$\kappa = \frac{\lambda_B A_A(\mu/\rho)_A^{BK\alpha}}{\lambda_A A_B(\mu/\rho)_B^{AK\alpha}}\sin\alpha \qquad (30)$$

in which λ_i is the wavelength of the K absorption edge; A_A and A_B are the atomic weights; and α is the take-off angle.

In order that eq. (28) may apply to a particular binary alloy system, it is necessary that a plot of C_A/K_A vs. C_A be a straight line with a slope of $(1 - a_{AB})$ and an intercept of α_{AB} at $C_A = 0$. This results from writing eq. (28) in the following form:

$$C_A/K_A = a_{AB} + (1 - a_{AB})C_A \qquad (31)$$

When working with a multicomponent system, it has been assumed that a_{AB} may be replaced by

$$\bar{a}_{At} = \sum_{B}^{t} a_{At}C_i/1-C_A \qquad (32)$$

i.e., the conversion parameter is a linear sum of the binary conversion parameters. The use of eq. (32) for ternary alloys has proved to work as well as the conventional method described by eq. (15). This statement is certainly borne out from the histograms published by Thomas[24].

Table I

Element A is Si	Line A is Si $K\alpha$	$\lambda_A = 7.126$	kV $= 30$	$\theta = 15.5°$
Element B is Nb	Line B is Nb $L\alpha$	$\lambda_B = 5.727$		csc $\theta = 3.74$

Absorption coefficients for line A		Absorption coefficients for line B	
$(\mu/p)_A^a = 327$	$\chi_A^a = 1220$	$(\mu/p)_A^b = 2323$	$\chi_A^b = 8680$
$(\mu/p)_B^a = 1334$	$\chi_B^a = 5170$	$(\mu/p)_B^b = 783$	$\chi_B^b = 2930$

C_A Si		1.000	0.800	0.600	0.377	0.167	0.154	0.092	0
C_B Nb		0	0.200	0.400	0.623	0.833	0.846	0.908	1.000
					Backscatter-ionization correction				
$R^* = R_AC_A + R_BC_B$	line A	0.935	0.908	0.881	0.851	0.823	0.821	0.812	0.810
	line B	0.940	0.914	0.888	0.859	0.832	0.830	0.822	0.810
$1/r^* = (1/r_A)C_A + (1/r_B)C_B$		0.43	0.404	0.378	0.349	0.322	0.320	0.312	0.30
					Absorption correction				
$Z^*_h = Z_AC_A + Z_BC_B$		14	19.4	24.8	30.8	36.5	36.9	38.6	41
		0.18	0.126	0.105	0.087	0.076	0.075	0.071	0.066
Line $A\sigma_c = 1477$									
$\chi^* = \chi_B^aC_A + \chi_B^aC_B$		1220	2010	2800	3680	4510	4560	4810	5170
$f(\chi)$		0.486	0.366	0.292	0.238	0.203	0.201	0.193	0.182
Line $B\sigma_c = 1488$									
$\chi^* = \chi_A^bC_A + \chi_B^bC_B$		8680	7530	6380	5100	3890	3815	3430	2930
$f(\chi)$		0.077	0.106	0.135	0.178	0.234	0.240	0.270	0.302
					Total correction				
Line $A(R^*/R_A)[(1/r_A)/(1/r^*)]$ =			1.033	1.071	1.120	1.18	1.18	1.20	
$f(\chi)AB/f(\chi)A$ =			0.755	0.581	0.491	0.418	0.414	0.398	
$1 + K_f$			1	1	1	1	1	1	
K_A/C_A =			0.781	0.622	0.550	0.493	0.489	0.478	
K_A =			0.625	0.373	0.207	0.082	0.075	0.044	
Line $B(R^*/R_B)(1/r_B)/(1/r^*))$ =			0.837	0.869	0.911	0.956	0.959	0.978	
$f(\chi)BA/f(\chi)B$ =			0.351	0.447	0.590	0.775	0.795	0.894	
$1 + K_f$			1	1	1	1	1	1	
K_B/C_B =			0.294	0.389	0.537	0.790	0.763	0.873	
K_B =			0.059	0.155	0.335	0.616	0.645	0.792	

3.1. Sample Calculation

The Cb–Si binary system has been selected as an example for several reasons: (1) there is no extensive solubility in any of the phases, (2) there is a large difference in the atomic numbers, and (3) both of the X-ray spectrum lines have high absorption effects.

Table I illustrates the calculation for the atomic number and absorption correction. The fluorescent correction in this case may be neglected. The results of these calculations, shown in figure 10, have been plotted in two ways, i.e., C/K vs. C and K vs. C. Experimental points are also shown that have been made on the intermetallic compounds and very fine-grained two-phase alloys. The two-phase alloys were analyzed with a large beam,

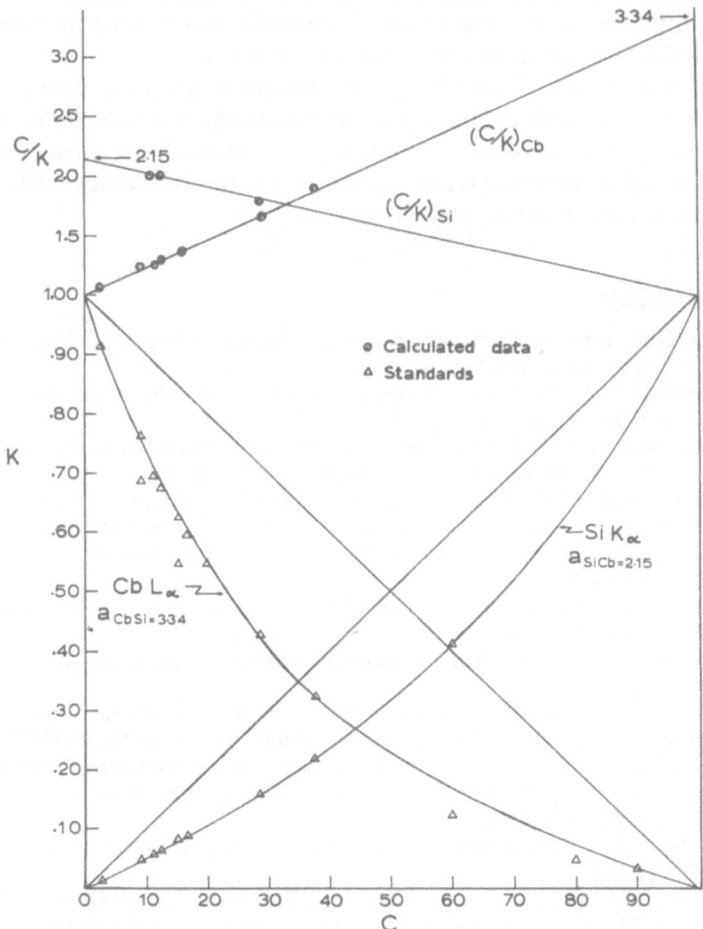

Figure 10. Calibration curves for the Cb–Si binary system.

and one cannot explain why they fall on the calibration curve. However, Moll[26] has shown this to be true for many two-phase systems.

One important feature of figure 10 is that the calculated points for C/K from Table I fall on a straight line. Only slight deviations from a straight line have been noticed when there is a large fluorescent correction as found in the Ni–Fe system, but even here the use of eq. (28) has proved to be the better procedure.

4. CONCLUSION

In spite of the cumbersome-looking equations, the data from the electron microprobe analyzer can be converted into chemistry. It is unsurpassed in its resolution, the range of elements that can be measured, and in particular the speed that the data can be taken.

We have assumed that the X-ray detecting system is linear, that the specimen remains in light optical focus during the measurements, and that the focusing circle of the X-ray optics passes through the X-ray source.

When all factors are taken into account, the instrument will provide information unequaled by any other technique.

REFERENCES

1. R. Castaing, Electron Probe Microanalysis *in* Advances in Electronics and Electron Physics, Vol. 13, pp. 317–386, Academic Press, New York (1960).
2. R. Castaing and J. Descamps, On the Physical Principles Underlying Point Analysis by X-ray Spectrography, *J. Phys. Rad.* **16**, 304–317 (1955).
3. P. Duncumb and P. K. Shields, The Present State of Quantitative X-ray Microanalysis Part 1: Physical Basis, *Brit. J. Appl. Phys.* **14**, 617–625 (1963).
4. G. D. Archard and T. Mulvey, The Present State of Quantitative X-ray Microanalysis Part 2: Computational Methods, *Brit. J. Appl. Phys.* **14**, 626–634 (1963).
5. H. G. Moseley, The High-Frequency Spectra of the Elements, *Phil. Mag.* **26**, 1024–1034 (1913).
6. R. Castaing and A. Guinier, Application of Electron Beams to Metallographical Analysis, *Proc. Intern. Conf. Electron Microscopy*, Delft, 60–63 (1949).
7. H. Kulenkampff, Über das kontinuierliche Röntgenspektrum *Ann. Physik* **69**, 548–596 (1922).
8. O. C. Wells, Electron Beams in Microelectronics *in* Introduction to Electron Beam Technology (R. Bakish, ed.), pp. 354–381, John Wiley, New York (1962).
9. D. B. Brown and R. E. Ogilvie, An Electron Transport Model for the Prediction of X-ray Production and Electron Backscattering in Electron Microanalysis, *J. Appl. Phys.* **37**, 4429–4433 (1966).
10. E. J. Sternglass, Backscattering of Kilovolt Electrons from Solids, *Phys. Rev.* **95**, 345–358 (1954).
11. R. E. Ogilvie and R. Lewis, Analysis of Al–Zn Concentration Gradients by Back-scattered Electrons, 8th Annual Denver Conference on Applications of X-ray Analysis (1959).

12. D. Wittry, An Electron Probe for Local Analysis by Means of X-ray, Ph.D. Thesis, California Institute of Technology, Pasadena, California (1957).
13. T. O. Ziebold, Diffusion Analysis of Phase Equilibra in Cu–Ag–Au Ternary Alloys, M.S. Thesis, Department of Metallurgy, Massachusetts Institute of Technology (June, 1963).
14. C. R. Worthington and S. G. Tomlin, The Intensity of Emission of Characteristic X-Radiation, *Proc. Phys. Soc.* (*London*) A69, 401–412 (1956).
15. G. D. Archard, Backscattering of Electrons, *J. Appl. Phys.* 32, 1505–1509 (1961).
16. M. Green and V. E. Cosslett, The Efficiency of Production of Characteristic X-Radiation in Thick Targets of a Pure Element, *Proc. Phys. Soc.* (*London*) 78, 1206–1214 (1961).
17. D. B. Brown and R. E. Ogilvie, Efficiency of Production of Characteristic X-radiation from Pure Elements Bombarded by Electrons, *J. Appl. Phys.* 35, No. 2, 309–314 (1964).
18. A. H. Compton and S. K. Allison, *X-rays in Theory and Experiment*, Van Nostrand, Princeton, New Jersey (1935).
19. D. L. Webster, K-Electron Ionization by Direct Impact of Cathode Rays, *Proc. Nat. Acad. Sci.* 14, 339–344 (1928).
20. D. M. Poole and P. M. Thomas, Quantitative Electron-Probe Microanalysis, *J. Inst. Metals* 90, 228–233 (1962).
21. P. Duncumb and P. K. Shields, Effect of Critical Excitation Potential on the Absorption Correction, *in* The Electron Microprobe (T. McKinley, K. Heinrich, and D. Wittry, eds.) pp. 284–295, John Wiley, New York (1966).
22. M. Green, A Monte Carlo Calculation of the Spatial Distribution of Characteristic X-ray Production in a Solid Target, *Proc. Phys. Soc.* (*London*) 82, 204–215 (1963).
23. I. Adler and J. I. Goldstein, *Absorption Tables for Electron Probe Microanalysis*, NASA Technical Note (1965).
24. P. M. Thomas, A Method for Correcting for Atomic Number Effects in Electron Probe Microanalysis, AERE Report, R4593 (1964).
25. T. O. Ziebold and R. E. Ogilvie, An Empirical Method for Electron Microanalysis, *Anal. Chem.* 36, 322–327 (1964).
26. S. Moll, Practical Methods for Experimentally Calibrated Concentration Determination in the Electron Beam Microanalyser, Proceedings of the First International Conference on Electron and Ion Beam Science and Technology (R. Bakish, ed.). p. 825, John Wiley, New York (1965).

IV. DETERMINATION OF SPECIFIC SURFACES BY SMALL-ANGLE X-RAY SCATTERING METHODS— A Brief Review

H. Brumberger

Department of Chemistry, Syracuse University, Syracuse, New York

The relationship between the specific surface (or more correctly the specific interfacial surface) of two-phase systems and the angular distribution of monochromatic X-rays scattered from such systems is shown on the basis of theories due to Debye and to Porod. The experimental technique (small-angle X-ray scattering) and the practical evaluation of observed scattering curves are discussed briefly. Measurements on solid porous catalysts, partly crystalline polymers, and other materials are cited to illustrate the applicability of this method to a wide variety of systems.

1. INTRODUCTION

The determination of specific surfaces is of particular significance to chemists concerned with adsorption effects (for instance, on electrode surfaces or in colloidal suspensions), heterogeneous catalysis in all its aspects, and partly crystalline polymers, gels, glasses, and similar systems. Many such measurements (especially on catalysts and other porous solids) have been made by adsorption isotherm methods, which are often tedious. The use of X-ray scattering may afford certain advantages, such as relative rapidity, nondestructiveness, and the ability to measure interfacial surfaces in systems where such surfaces may not be accessible to adsorbed gases. In addition, the X-ray data may simultaneously yield structural information.

In this article, the discussion is confined to systems where the dimensions of the regions of interest are in the colloidal range, i.e., from some tens to some thousands of angstroms; this includes many of the important cases. If one thinks for a moment of Bragg's law, it is evident from the inverse relation between spacing d and scattering angle 2θ at fixed wavelength λ that one must go to small angles for observations related to large structural features. One consequently deals here with small-angle X-ray

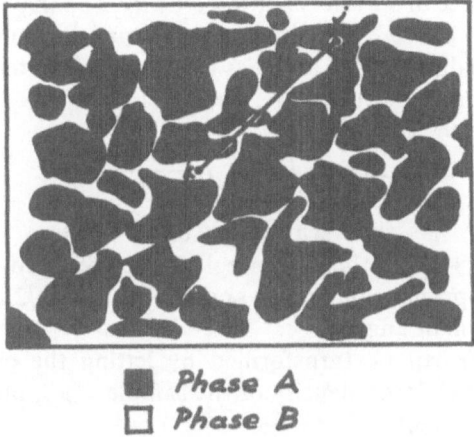

■ Phase A
□ Phase B

Figure 1. The two-phase model.

scattering, and the measurements are conditioned by the requirements of this technique. The scattering, then, is caused by electron density inhomogeneities of colloidal dimensions.

This paper largely restricts discussion of the theory to a simple model which appears to work reasonably well with many experimental samples, i.e., that of a two-phase system containing regions of phase A and uniform electron density ρ_A and regions of phase B and uniform electron density ρ_B (figure 1). The phase boundaries between A and B regions are considered sharp. This picture can be applied with equal appropriateness to a dilute colloidal suspension of, for instance, latex spheres in water, a densely packed solid aggregate of porous alumina, or to a solid polymer containing crystalline and amorphous regions. Densely packed systems usually provide the greatest difficulty in surface measurements and the remarks following Section 2 are primarily intended to apply to such systems. Other methods (i.e., electron microscopy) are available for dilute systems.

2. GENERAL THEORETICAL BACKGROUND

If the location of a scattering volume element j of the sample, possessing an electron density ρ_j, is given by the vector $\mathbf{x}_j$, then the intensity I scattered from the illuminated volume V at an angle 2θ to the primary X-ray beam is, according to classical scattering theory[1,2],

$$I(\mathbf{h}) = I_e(h) \int_V \int_V \rho(\mathbf{x}_j)\rho(\mathbf{x}_k) e^{-ih.(x_j - x_k)} dx_j \, dx_k \qquad (1)$$

$\rho d\mathbf{x}$ is the equivalent of the scattering factor, $e^{-ih \cdot (x_j - x_k)}$ takes into account the phase difference of wavelets scattered by dx_j and dx_k in the 2θ-direction, and $|\mathbf{h}| = 4\pi\lambda^{-1} \sin \theta$. The integrals add up the contributions to the scattering in this direction of all pairs of scattering centers in V. $I_e(h)$ is the scattering by a single electron, i.e.,

$$I_e(h) = \left(\frac{e^2}{mc^2}\right)^2 I_0 \frac{1 + \cos^2 2\theta}{2R^2} \tag{2}$$

where $(e^2/mc^2)^2$ is the scattering cross section of the electron (7.9×10^{-26} cm^2); R is the distance from scatterer to detector in centimeters; I_0 is the primary intensity; and $(1 + \cos^2 2\theta)/2$ is the polarization factor (unity at small angles).

Equation (1) can be transformed by letting the electron density ρ_j equal the average electron density of the sample, $\langle \rho \rangle$, plus a local electron density fluctuation $\delta\rho_j$;

$$\rho_j = \langle \rho \rangle + \delta\rho_j \tag{3}$$

and by setting $\mathbf{x}_k = \mathbf{x}_j + \mathbf{r}$. Then it may be shown[2] that

$$I(\mathbf{h}) = I_e(h) \int_{v'} \int_v \delta\rho(\mathbf{x}_j)\delta\rho(\mathbf{x}_j + \mathbf{r}) e^{i\mathbf{h}\cdot\mathbf{r}} dx_j \, d\mathbf{r} \tag{4}$$

is the only significant term left in eq. (1) at any experimentally accessible scattering angle. One may now define a function $G(\mathbf{r})$, called correlation function by Debye[3] and characteristic function by Porod[4]:

$$\langle(\delta\rho)^2\rangle G(\mathbf{r}) = \frac{1}{V} \int_V \delta\rho(\mathbf{x}_j)\delta\rho(\mathbf{x}_j + \mathbf{r}) \, dx_j \tag{5}$$

or

$$G(\mathbf{r}) = \langle \delta\rho_j \delta\rho_k \rangle / \langle(\delta\rho)^2\rangle \tag{6}$$

Thus,

$$I(\mathbf{h}) = I_e(h) \langle(\delta\rho)^2\rangle V \int_V G(\mathbf{r}) e^{i\mathbf{h}\cdot\mathbf{r}} \, d\mathbf{r} \tag{7}$$

For a spatially isotropic medium, $G(\mathbf{r}) = G(r)$; after integration over the angular variables, eq. (7) then reduces to

$$I(h) = I_e(h) \langle(\delta\rho)^2\rangle V \int_0^\infty 4\pi r^2 G(r) \frac{\sin hr}{hr} dr \tag{8}$$

where the upper limit can be set at infinity since $G(r) \to 0$ for large r: $\langle \delta\rho_j \delta\rho_k \rangle \to \langle \delta\rho_j \rangle \langle \delta\rho_k \rangle = 0$ if $\delta\rho_j$ and $\delta\rho_k$ are independent, as they would be for $r \to \infty$.

$\langle(\delta\rho)^2\rangle G(r)$ is now the value of the product of electron density fluctuations at the two ends of r, averaged as r moves around the medium taking on all possible positions or orientations but maintaining a constant length.

From eq. (8) it is apparent that $rG(r)$ and $hI(h)$ are a pair of Fourier transforms; $G(r)$ contains the essential structural information obtainable from a scattering experiment. $G(r)$ may be determined by numerical Fourier inversion of the experimental intensity data—an undertaking subject to considerable errors—or may be calculated on the basis of a specific structural model and the resulting angular intensity distribution from eq. (8) compared to observation.

The task which remains is to establish a connection between specific surface and $G(r)$. A specific model is needed, and it is here that one introduces the two-phase concept described in Section 1.

3. CORRELATION FUNCTION AND SPECIFIC SURFACE

If only two phases are present—for instance, solid and pores—the ends of $\mathbf{r}$ (the vector connecting dx_j and dx_k) may be in only two environments, A or B, and only four situations are possible—AA, AB, BA, and BB. If one calculates for a given r in an isotropic medium the probability P_D of having different environments for the two ends[3], one finds

$$P_D = 2\phi_A(1 - \phi_A)[1 - G(r)] \tag{9}$$

where ϕ_A is the volume fraction of phase A. P_D may be related to the surface-to-volume ratio by the following reasoning[3]: Supposing one requires r to become very small, to take on all possible orientations in space, and to have its ends in different phases and inquires into the resulting expression for P_D. Assume that j is in phase A, in the interval dl, a perpendicular distance l from the interface; $l \leqslant r$, $r \to 0$. The probability of finding j in this location, remembering that l is very small, is Sdl/V, where S is the total interfacial area; Sdl is the favorable volume; and V is the total volume of the sample. While j is in a position which allows k to be in the other phase, r may take on all possible orientations. The locus of possible locations for k is the surface of a sphere of radius r which is either tangent to the interface or intersects it. The probability that k is in phase B is equal to the fraction of the sphere's surface protruding above the interface into B, i.e., $2\pi r(r - l)/4\pi r^2$. The product of these two factors gives the probability that j and k are simultaneously in different environments for a given distance l of j from the interface. To get the total probability, we must integrate this product over all values of l which allow j and k to be dissimilar and multiply by 2 to account for the situation where j is in phase B to begin with; thus,

$$P_D = 2 \int_0^r \frac{S dl}{V} \cdot \frac{2\pi r(r - l)}{4\pi r^2} = \frac{Sr}{2V} = 2\phi_A(1 - \phi_A)[1 - G(r)] \qquad (10)$$

Differentiating with respect to r and evaluating the derivative at $r = 0$ [a precondition for eq. (10)], we obtain

$$\frac{S}{V} = -4\phi_A(1 - \phi_A)G'(0) \qquad (11)$$

The surface-to-volume ratio of a material whose volume fractions are known thus may be calculated from the limiting value of the first derivative of its correlation function, obtainable from X-ray measurements.

In the case of a porous solid with a random distribution of solid regions, it was shown by Debye, Anderson, and Brumberger[3] that

$$G(r) = e^{-r/a} \qquad (12)$$

where a is a correlation length characteristic of the material. When eq. (12) applies, eqs. (8) and (11) reduce to eqs. (8a) and (11a):

$$I(h) = \frac{I_e(h) \cdot 8\pi a^3 \langle (\delta\rho)^2 \rangle V}{(1 + h^2 a^2)^2} \qquad (8a)$$

and

$$\frac{S}{V} = \frac{4\phi_A(1 - \phi_A)}{a} \qquad (11a)$$

where a can be determined from the slope–intercept ratio of a plot of $I(h)^{-1/2}$ vs. h^2, and S/V is then immediately accessible via eq. (11a).

An alternative method for determining S is afforded by a measurement of the absolute scattered intensity, i.e., of $I(h)/I_0$. It may be shown[3] that $I(h)$ can be developed as a series expansion in $1/h$, and that at sufficiently large angles (or values of h) the major contribution to the intensity is made by the leading term, which has the form

$$\frac{I(h)}{I_e(h)} = \frac{-8\pi \langle (\delta\rho)^2 \rangle V G'(0)}{h^4} \qquad (13)$$

Thus, our two-phase model should show a scattered intensity proportional to $1/h^4$ or, for some range of angles, $h^4 I(h)$ should reach a limiting constant value k.

From eq. (11) it then follows that

$$S = \frac{\phi_A(1 - \phi_A)k}{2\pi I_e(h) \langle (\delta\rho)^2 \rangle} \qquad (14)$$

Since the mean square electron density fluctuation for the sample is given by

$$\langle (\delta\rho)^2 \rangle = \phi_A(1 - \phi_A)(\Delta\rho)^2 \tag{15}$$

where $\Delta\rho$ is the electron density difference between the two phases, $\rho_A - \rho_B$, one finally obtains

$$S = \frac{k}{2\pi I_e(h)(\Delta\rho)^2} \tag{16}$$

For a fixed incident intensity, sample thickness, and material, the measurement of intensity at one angle in the region where $I(h) \propto 1/h^4$ for a series of samples will then immediately establish relative surface areas; only one calibrated sample is needed to convert these to absolute areas[5].

The measurement of I_0, required to apply eq. (16) for determining absolute surfaces, may be made by means of a rotating sector method[6], by using calibrated attenuators[7], by measuring the scattering of a gas whose composition and equation of state are accurately known[8], or by using a standard comparison sample calibrated by one of these methods[9].

It is possible to circumvent the absolute intensity measurement if the volume fractions of the phases are known and it can be established that the scattering curve indeed shows a slope of -4 in some region of a plot of $\log I(h)$ vs. $\log h$. Porod[4] showed that the quantity Q, called the *invariant* and defined by

$$Q = \int_0^\infty I(h)h^2 \, dh \tag{17}$$

can be related to the mean-square electron-density fluctuation $\langle (\delta\rho)^2 \rangle$ by the equation

$$Q = 2\pi^2 I_e(h) V\phi_A(1 - \phi_A)(\Delta\rho)^2 \tag{18}$$

Consequently, one may derive without difficulty an expression for S/V from eq. (16):

$$\frac{S}{V} = \frac{k\pi\phi_A(1 - \phi_A)}{Q} \tag{19}$$

The $I_e(h)$ term has been eliminated, and the measurement of S/V now involves calculating the area under the curve of $h^2I(h)$ obtained experimentally and taking the ratio of $k\pi\phi_A(1 - \phi_A)$ to this area. It is generally easiest to plot $h^4I(h)$ vs. h^4 and determine the value of k graphically by extrapolation of the constant portion of this curve to $h = 0$. The curve $I(h)h^2$ must begin at the origin and generally will increase and then decrease again as $I(h)$ becomes small at larger angles. With strong scatterers, it may be difficult to extrapolate the small-angle part of $h^2I(h)$ to the origin since the curve may continue to rise through the entire angular range accessible experimentally.

4. EXPERIMENTAL TECHNIQUE

Since it is desirable to make measurements at small scattering angles, an X-ray camera with a very good collimating system and long sample-to-detector path is required; the latter should be evacuated to avoid air scattering. A beam stop is needed to prevent the direct beam from reaching the plate when making a photographic record of sample scattering. Most modern cameras, however, rely on counter detection. Several alternative methods for monochromatizing the X-radiation are available—the use of balanced Ross filters[10], the combination of a $K\beta$-filter, a proportional counter and a pulse height discriminator, crystal monochromatization, etc. It is essential to stabilize the X-ray source output with all counter–detector methods or to monitor the output with a second counter. Since much of this work is done with very narrow slits to improve angular resolution, it is also important to stabilize the position of the focal spot of the X-ray tube relative to the camera collimator. To this end, the cooling water of the X-ray unit should be maintained at constant temperature, and the room in which the unit is located should be air-conditioned.

The use of slits rather than pinholes to define the geometry of the X-ray beam imposes a distortion on the intensity pattern; the slit acts like a rectangular array of pinholes, each contributing a portion of the total observed result. The calculation of undistorted intensity distributions from the observed pattern and a knowledge of the slit geometry is generally an extremely difficult problem; however, solutions are available for the most important special cases[11] and various computational methods have been developed for their calculation[12,13,14]. In many cases it is possible to work with the slit-smeared observed intensities directly. For example, one may show that the theoretical $1/h^4$-dependence of the intensity scattered by a two-phase system at larger scattering angles for perfect pinhole collimation reduces to a dependence on $1/h^3$ if the primary beam profile at the detector may be considered "infinitely" long and negligibly wide relative to the angular widths of the small-angle pattern. If we consider figure 2b, $P(t)$ is the intensity distribution for just such a beam profile; the uniform constant portion is sufficiently long relative to the angular width of the observed pattern so that end effects are negligible (criteria for this will be found in Kratky et al.[11]). The width (in the m direction, where m is a micrometer reading for the angular motion of the detector slit and is equal to $h\lambda R/2\pi$) is very small (figure 2a). Each portion dt of the beam contributes an

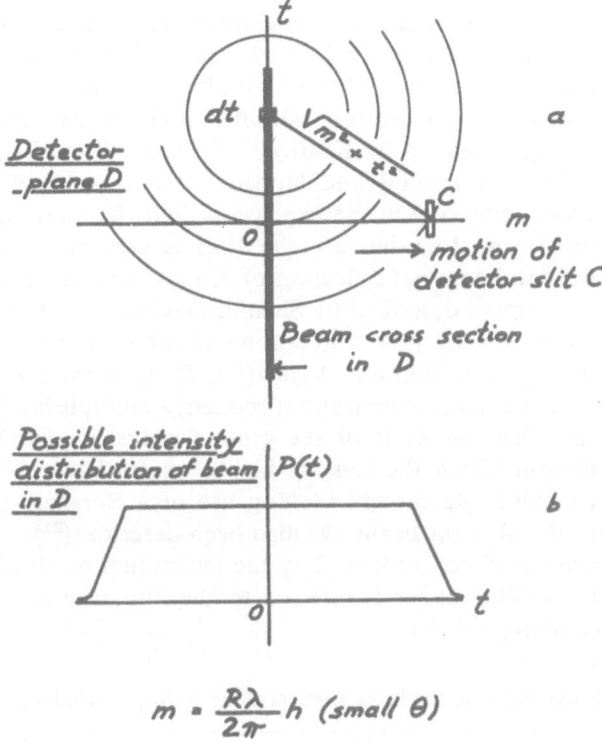

$$m = \frac{R\lambda}{2\pi}h \; (small \; \theta)$$

Figure 2. Slit corrections for an "infinitely" long primary beam.

intensity to the detector slit (itself considered negligibly small compared to the beam dimensions) which is proportional to $P(t)dt$ and to the intensity $I(\sqrt{m^2 + t^2})$ scattered at an angle corresponding to the distance $\sqrt{m^2 + t^2}$ by the scatterer illuminated by a beam of very small cross section. Then the observed intensity $\tilde{I}(m)$ is the sum of all such contributions:

$$\tilde{I}(m) = 2\int_0^\infty P(t)I(\sqrt{m^2 + t^2}) \, dt \tag{20}$$

with $P(t)$ equal to a constant which can be set equal to unity for the "infinite" case. The calculated I then will be normalized to an intensity per unit length of primary beam. From eq. (20), if I has a $1/h^4$ dependence, $\tilde{I}$ can be shown to be proportional to $1/h^3$.

A number of small-angle cameras have been designed. Among these are instruments based on conventional slit systems, focusing cameras, cameras with unconventional collimating systems, etc.

A widely used type of instrument employs the four-slit geometry of Beeman[15], with two slits between X-ray source and sample and two more between sample and detector; the third slit eliminates parasitic scattering. A different camera design permitting extremely high angular resolution was developed and highly refined by Kratky[16,17]; the collimator in this unit consists of a series of ground and lapped coplanar steel blocks which eliminate parasitic scattering at the expense of half the scattering pattern. Instruments employing focusing are the Franks camera[18], which uses optical flats and total external reflection of X-rays, and the double crystal monochromator camera described by Shenfil, Davidson, and DuMond[19]. An entirely different concept for obtaining ultrahigh angular resolution was recently described by Bonse and Hart[20]. Their device uses a grooved single crystal between X-ray source and specimen; a multiple Bragg reflection taking place between the walls of the groove yields a relatively intense rocking curve from which the tails at higher angles have been virtually eliminated. A small-angle camera making use of a Borrman germanium crystal to form the incident beam has also been described[21].

Some specific surfaces measured by the techniques outlined above are shown in Table I. The reader is referred to the literature for details and further applications ([3, 22-28]).

Table I. Some Specific Surfaces Measured by X-Ray Techniques (m^2/g)

Substance	Specific surface		Reference
	X-ray	BET	
Al_2O_3	255	210	([3])
Al_2O_3	506	399	([3])
Al_2O_3	145	116	([3])
Al_2O_3–SiO_2	396	254	([3])
Montmorillonite	117		([28])
SiO_2	206	310	([28])
Nylon 6	176		([22])
Charcoal	27.5		([28])
Ni powder	3.56		([28])

REFERENCES

1. A. Guinier, *X-Ray Diffraction*, W. H. Freeman, San Francisco (1963).
2. A. Guinier *et al.*, *Small-Angle Scattering of X-Rays*, J. Wiley, New York (1955).
3. P. Debye, H. R. Anderson, and H. Brumberger, *J. Appl. Phys.* **28**, 679 (1957).
4. G. Porod, *Kolloid-Z.* **124**, 83 (1951).

5. R. A. Van Nordstrand and K. M. Hach, American Chemical Society Meeting, Chicago (September, 1953).
6. O. Kratky and H. Wawra, *Monatsh. Chem.* **94**, 981 (1963).
7. V. Luzzati, *Acta Cryst.* **13**, 939 (1960).
8. W. W. Beeman, *Proceedings of the Syracuse Conference on Small-Angle X-Ray Scattering*, Gordon and Breach, New York (1967).
9. O. Kratky, I. Pilz, and P. J. Schmitz, *J. Coll. Interface Sci.* **21**, 24 (1966).
10. R. A. Young, *Z. Krist.* **118**, 233 (1963).
11. O. Kratky, G. Porod and Z. Skala, *Acta Phys. Austriaca* **13**, 76 (1960).
12. P. W. Schmidt and R. Hight, *Acta Cryst.* **13**, 480 (1960).
13. J. Mazur and A. M. Wims, *J. Res. Natl. Bur. Stand.* **70A**, 467 (1966).
14. S. Heine and J. Roppert, *Acta Phys. Austriaca* **15**, 148 (1962).
15. J. W. Anderegg, W. W. Beeman, and S. Shulman, *Phys. Rev.* **87**, 186 (1952).
16. O. Kratky and Z. Skala, *Z. Elektrochem.* **62**, 73 (1958).
17. O. Kratky, *Progr. Biophys.* **13**, 105 (1963).
18. A. Franks, *Proc. Phys. Soc.* **68B**, 1054 (1955).
19. L. Shenfil, W. E. Danielson, and J. W. M. DuMond, *J. Appl. Phys.* **23**, 854 (1952).
20. U. Bonse and M. Hart, *Appl. Phys. Letters* **7**, 238 (1965).
21. H. Brumberger and R. Deslattes, *J. Res. Natl. Bur. Stand.* **68C**, 173 (1964).
22. H. Brumberger, O. Kratky, and P. Mittelbach, *Monatsh. Chem.* **95**, 1599 (1964).
23. D. Weigel, A. Renouprez, and B. Imelik, *J. Chim. Phys.* **62**, 125 (1965).
24. A. Renouprez, H. Bottazzi, D. Weigel, and B. Imelik, *J. Chim. Phys.* **62**, 131 (1965).
25. O. Kratky and K. Schwarzkopf-Schier, *Monatsh. Chem.* **94**, 714 (1963).
26. P. H. Hermans and A. Weidinger, *Makromol. Chem.* **39**, 67 (1960).
27. G. Porod, *Makromol. Chem.* **35**, 1 (1960).
28. L. Kahovec, G. Porod, and H. Ruck, *Kolloid-Z.* **133**, 16 (1953).

V. DETERMINATION OF PARTICLE-DIAMETER DISTRIBUTIONS BY SMALL-ANGLE X-RAY SCATTERING*

Paul W. Schmidt

and

Charles G. Weil†

*Physics Department, University of Missouri
Columbia, Missouri*

and

Orville L. Brill

*Physics Department, Kansas State College
Pittsburg, Kansas*

Small-angle X-ray scattering can be used to obtain information about the distribution of particle diameters in dilute polydisperse suspensions of colloidal particles with diameters in a range from about 30 to 1000 Å. The usual analysis of the complete small-angle scattering curve from a dilute colloidal suspension gives three parameters characterizing the suspended particles in the sample: specific surface, volume-average volume, and radius of gyration. From each of these quantities an average particle diameter can be calculated. Since each of the three average diameters is a differently weighted average, comparison of these diameters provides qualitative information about the particle-diameter distribution. For spherical particles, diameter distributions can be directly calculated from the scattering data with almost no assumptions about the form of the diameter distribution. The theory of this technique is outlined, and its use is illustrated by a summary of a calculation of particle-diameter distributions from experimental scattering curves obtained for three silica suspensions.

1. INTRODUCTION

Small-angle X-ray scattering, which is a well-known method for studying the form and average dimensions of colloidal particles, can also give

* Work supported by the National Science Foundation.

† National Science Foundation undergraduate research assistant. Current address: Physics Department, University of Illinois, Urbana, Illinois.

information about the distribution of the diameters of the particles in polydisperse colloidal suspensions. The X-ray method for investigating particle diameter distributions has the important advantage of being well suited for particles with diameters in the range from about 30 to 1000 Å. For such small particles the only other common technique for finding diameter distributions is to measure particle diameters on electron micrographs. These measurements are difficult or impossible when the average diameter is less than about 100 Å. Because of the scarcity of other ways of determining particle-diameter distributions in suspensions of relatively small colloidal particles, small-angle X-ray scattering is an important method of studying and characterizing these suspensions.

Two different procedures are discussed here. The first method, which gives qualitative and semiquantitative information about diameter distributions, consists of comparing the different average particle dimensions which are obtained in the usual analysis of small-angle X-ray scattering data. This comparison provides information about the diameter distribution, since each average dimension represents a differently weighted average. The second method, which is applicable only to suspensions of spherical particles, permits calculation of the diameter distribution function from the scattering data while requiring almost no assumptions about the properties or form of this function.

2. ASSUMPTIONS AND RANGE OF APPLICABILITY

In the interpretation of the X-ray data, the colloidal samples will be considered to be so dilute that in the interval of scattering angles for which data are available the scattering curve is unaffected by interparticle interactions. With this assumption, the observed scattering is proportional to the intensity scattered by a single particle. In addition, the particles will be assumed to be randomly oriented, so that the measured scattering is the average over all particle orientations.

The assumption of independent, randomly oriented particles can usually be satisfied by making the sample sufficiently dilute. In most cases, the upper concentration limit will be of the order of 1 to 5 % by weight. In experimental studies, the independence of the particles can be tested by measuring the scattering at several concentrations. For concentrations low enough that the intensity at a given scattering angle is proportional to the concentration, the scattering can ordinarily be considered to be due to independent, randomly oriented particles.

Also, the particles will be assumed to have a constant electron density and to be suspended in a fluid in which the electron density has a different, but also constant, value. Although this assumption neglects the atomic and molecular structure of the colloidal particles and the presence of a diffuse electrical double layer, it can be considered to be well satisfied in small-angle X-ray scattering[1,p.4].

To simplify the interpretation of the data, the polydisperse suspension will be assumed to be composed of particles with the same shape. This assumption means that the form and dimensions of a particle are completely determined when one particle dimension is specified. The dimension which will be used in this discussion is the longest straight line that can be contained inside the particle. This length is called the maximum diameter a.

With these assumptions, the polydisperse colloidal suspension can be described by a diameter distribution function $\rho(a)$, which is defined to be a function such that $\rho(a)\,da$ represents the probability that the observed radiation has been scattered by a particle with a maximum diameter in a range between a and $(a + da)$. The diameter distribution function will be assumed to satisfy the normalization condition

$$1 = \int_0^\infty \rho(a)\,da$$

The scattering data consist of measurements of the scattered intensity as a function of the scattering angle θ. Since experimental methods are outlined in the paper by H. Brumberger, they will not be repeated here. Further details can be found in Guinier et al.[1]. The scattering angle is conveniently expressed by the quantity $h = 4\pi\lambda^{-1}\sin(\theta/2)$, where λ is the X-ray wavelength. The intensity will be denoted by the quantity $I(h)$. A scattering curve thus represents measurements of $I(h)$ as a function of h. For particles with diameters from 30 to 1000 Å, the scattering angles of interest range from a few minutes to a few degrees.

The values of $I(h)$ are assumed to be corrected for all errors and distortions produced by the experimental equipment.

3. COMPARISON OF AVERAGE DIMENSIONS

From the scattering curve for a dilute polydisperse suspension which satisfies the above conditions, the ratio of the number average particle surface area S to the number average particle volume V can be obtained from the scattering data by use of the relation[2]

$$S/V = \frac{\pi \lim_{h\to\infty} h^4 I(h)}{\int_0^\infty dh\, h^2 I(h)} \qquad (1)$$

In eq. (1) and in the equations below, bars above quantities indicate values for a single particle, averaged over the total number of particles in the colloidal suspension.

Also, where $\overline{V^2}$ is the average of the square of the volume of a particle[2],

$$\overline{V^2}/\overline{V} = \frac{2\pi^2 I(0)}{\int_0^\infty dh\, h^2 I(h)} \qquad (2)$$

At small h, $I(h)$ has the approximate form

$$I(h) = I(0)\exp(-\tfrac{1}{3}h^2\overline{R^2}) \qquad (3)$$

where R is the radius of gyration of a particle, and

$$\overline{R^2} = \overline{(V^2 R^2)}/(\overline{V^2}) \qquad (4)$$

From eqs. (1), (2), and (3), the quantities $\overline{V^2}/\overline{V}$, $\overline{S}/\overline{V}$, and $\overline{R^2}$ can be determined, provided the right side of each equation can be evaluated. For colloidal suspensions satisfying the above assumptions, small-angle X-ray scattering theory predicts that

$$\lim_{h \to \infty} h^4 I(h)$$

is a finite positive constant. In a number of experimental studies of colloidal suspensions, $h^4 I(h)$ has been found to be at least approximately constant in the outer part of the scattering curve. However, other results show that the behavior of $h^4 I(h)$ in the accessible range of scattering angles cannot be taken for granted but must be tested in each sample which is studied. But when data are available showing that in the outer part of the scattering curve the quantity $h^4 I(h)$ is essentially constant, $I(h)$ can be assumed to be proportional to h^{-4} for all angles larger than those for which data are available. With this assumption the $I(h)$ curve can be extrapolated to arbitrarily large h values, thus permitting numerical evaluation of the integral in the denominators of the right sides of eqs. (1) and (2).

For sufficiently small h, eq. (3) can be shown to hold for all suspensions of independent particles.* The theory, however, does not guarantee that this relation will be satisfied in the experimentally accessible region of scattering angles. The validity of eq. (3) for a given sample and angular region can be tested by making a plot, often called a radius of gyration plot or Guinier plot, of log $I(h)$ as a function of h^2. A straight-line plot indicates that eq. (3) is satisfied. From the slope of the line, $\overline{R^2}$ can be calculated. By extrapolation of the line to zero h, $I(0)$ can be found for use in the

---

* See Guinier *et al.*[1] pp. 24–27.

numerator of eq. (2), and the integrand in the denominators of eqs. (1) and (2) can be evaluated in the neighborhood of $h = 0$.

Thus, without any assumptions about the size and shape of the particles, direct analysis of the scattering curve gives $S/\bar{V}$, $\overline{V^2}/\bar{V}$, and $\overline{R^2}$. The procedure described below can be used to compute average particle dimensions.

When all particles are assumed to have the same shape, the volume $V(a)$ and surface area $S(a)$ of a particle with maximum diameter a can conveniently be written $V(a) = V_0 a^3$ and $S(a) = S_0 a^2$, where V_0 and S_0 are dimensionless quantities which have the same value for all particles of the sample and which can be calculated once the particle shape is known. Sufficient information about the suspension will be assumed to be available to permit a selection of a reasonable particle shape.

In terms of the diameter distribution $\rho(a)$, the quantities $S/\bar{V}$, $\overline{V^2}/\bar{V}$, and $\overline{R^2}$ can be written

$$\frac{S}{\bar{V}} = \frac{S_0 \int_0^\infty da\, a^2 \rho(a)}{V_0 \int_0^\infty da\, a^3 \rho(a)} \tag{5}$$

$$\frac{\overline{V^2}}{\bar{V}} = \frac{V_0 \int_0^\infty da\, a^6 \rho(a)}{\int_0^\infty da\, a^3 \rho(a)} \tag{6}$$

$$\overline{R^2} = (R_0)^2 \frac{\int_0^\infty da\, a^8 \rho(a)}{\int_0^\infty da\, a^6 \rho(a)} \tag{7}$$

where, for a particle with maximum diameter a and radius of gyration $R(a)$, $R_0 = [R(a)/a]$. Three differently weighted average diameters a_S, a_V, and a_R then can be computed by the equations

$$a_S = \frac{\int_0^\infty da\, a^3 \rho(a)}{\int_0^\infty da\, a^2 \rho(a)} = \frac{S_0}{V_0} \frac{\bar{V}}{S} \tag{8}$$

$$a_V = \left[\frac{\displaystyle\int_0^\infty da\, a^6 \rho(a)}{\displaystyle\int_0^\infty da\, a^3 \rho(a)} \right]^{\frac{1}{3}} = \left[\frac{1}{V_0} \frac{\overline{V^2}}{\overline{V}} \right]^{\frac{1}{3}} \tag{9}$$

$$a_R = \left[\frac{\displaystyle\int_0^\infty da\, a^8 \rho(a)}{\displaystyle\int_0^\infty da\, a^6 \rho(a)} \right]^{\frac{1}{2}} = (R_0)^{-1} (\overline{R^2})^{\frac{1}{2}}. \tag{10}$$

Under quite general conditions([4]), $a_R \geqslant a_V \geqslant a_S$.

In monodisperse suspension, $a_R = a_V = a_S$. In a polydisperse sample, these three quantities will be more nearly equal to each other for a narrow diameter distribution than when the distribution is broad.

For polydisperse suspensions the value of $S/\overline{V}$ calculated from the small-angle X-ray scattering data ordinarily is comparable to the specific surface determined from BET adsorption measurements. Within experimental uncertainty, the average particle diameter a_S therefore can be expected to agree with the average dimension calculated from BET data.

Similarly, since $V^2/\overline{V}$ is a volume average particle volume (equivalent to a weight average weight), a_V is an average diameter weighted in the same way as the average dimension calculated from light-scattering measurements.

Scattering curves were measured for three different colloidal silica suspensions: Sample I, specially prepared Ludox HS® in which the particle diameter distribution was narrower than that of Sample II, the commercial grade Ludox HS®, and Sample III, commercial grade Ludox SM®.

The scattering curves for the three samples are presented in figure 1. Table I lists the various average particle diameters computed from the scattering curves, assuming that the particles are spheres. As expected, the values of the three different average particle diameters differ less for Sample I than for Sample II, reflecting the narrower particle diameter distribution in Sample I. The scattering data thus indicate that even though the values of a_S are nearly the same in Samples I and II, $\rho(a)$ is considerably narrower in Sample I. As discussed in Section 4.3, this conclusion is verified by measurements of particle diameters on electron micrographs and by a more detailed analysis of the scattering data.

---

* ® Trademark E. I. du Pont de Nemours & Co., Inc. Sample I was obtained from the du Pont Laboratories, Wilmington, Delaware, and the other two samples were supplied by J. Kratohvil, Clarkson College of Technology, Potsdam, New York.

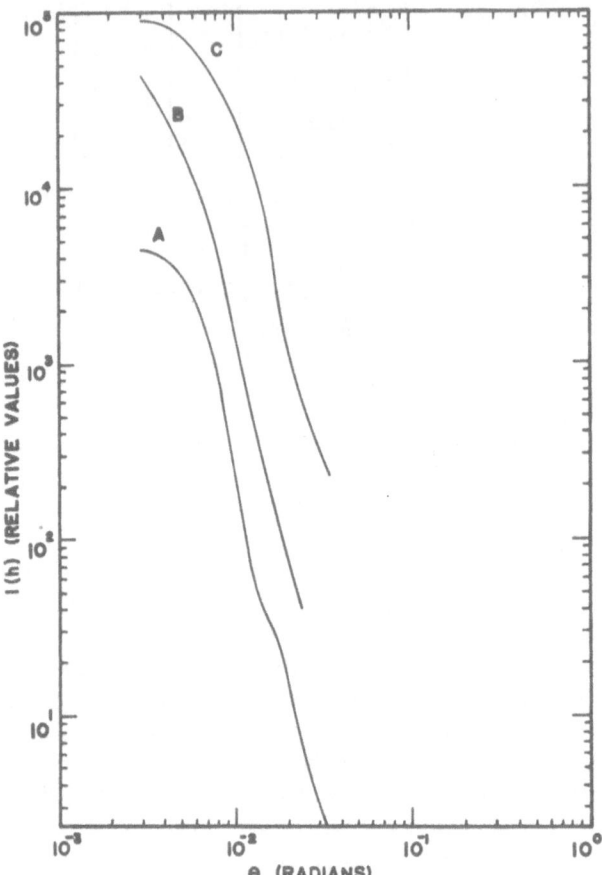

Figure 1. The relative scattered intensity for Samples I
(curve *A*), II (curve *B*), and III (curve *C*). For convenience
in plotting, the ordinates of the curves have been arbitrarily shifted with respect to each other. The curves,
which were obtained with X-rays with wavelength 1.54 Å,
have been corrected for the effects of the collimating slits.

Table I. Average Particle Diameters (Å)

Sample	a_S	a_V	a_R
I	141	167	180
II	142	218	246
III	111	117	124

For Sample III (Ludox SM®) the computed average diameter a_S is smaller than that in the other two samples. The other average diameters for this sample are also listed in table I.

In principle, a reasonable form can be assumed for $\rho(a)$, with certain parameters, including the width of the diameter distribution, being variable. These parameters then can be adjusted to provide a diameter distribution which most nearly gives all three average diameters obtainable from the scattering curve. Unfortunately, the usefulness of this procedure is limited by the uncertainties in the average diameters, which can be expected to be of the order of 10 to 20% because of uncertainties in the scattering data and because the samples do not perfectly satisfy the assumptions made in obtaining eqs. (1), (2), and (3). These uncertainties usually limit the quantitative information about $\rho(a)$ which can be obtained by comparison of the average dimensions, since the differences between the different average diameters are often of the same magnitude as the uncertainties in their values.

Mittelbach[5] assumed that $\rho(a)$ had the form

$$\rho(a) = [(n + 1)/a_0][(n - 3)!]^{-1}[(n + 1)(a/a_0)]^{n-3} \exp[-(n + 1)(a/a_0)]$$

where n is an integer and a_0 is a constant. With this diameter distribution he computed theoretical scattering curves for spherical particles for different values of n. By comparing experimental curves with these calculated curves the values of n and a_0 and the form of $\rho(a)$ can be estimated. However, this method cannot distinguish between slightly different forms of $\rho(a)$, since the theoretical curves are not very sensitive to small changes in n.

4. CALCULATION OF DIAMETER DISTRIBUTIONS

4.1. Introduction

A number of years ago Roess[6] and Riseman[7], working independently, showed that in principle the diameter distribution $\rho(a)$ could be calculated from the scattering data for a suspension of independent spherical particles. At the time this work was published, small angle X-ray scattering was not sufficiently developed to permit meaningful determinations of $\rho(a)$ from experimental scattering data, and this method was therefore considered to be of little importance.

For the past several years we have been reexamining the problem of calculating $\rho(a)$ from the scattering data. Since the practicality of the method had never been tested, our first studies[8] were concerned with two subjects. First, the general theoretical equations were modified to make them more convenient to apply to experimental scattering curves. This rearrangement permitted $\rho(a)$ to be expressed as an integral transform of a function which could be readily evaluated from the scattering data. Our other area of interest was the effect that errors in the scattering data can have on the computed diameter distribution. This effect was studied[8,9] by calculating $\rho(a)$ from theoretical scattering curves for which $\rho(a)$ could be found both analytically and by numerical evaluation of the integral transform. Diameter distributions were computed both from exact theoretical scattering curves and from theoretical curves modified to simulate the errors and uncertainties likely to be present in actual data. From these tests a procedure has been developed[9] which we feel is a practical way of calculating $\rho(a)$ from scattering curves for polydisperse suspensions of spherical particles. The theory has also been extended to particles with other shapes[10].

To test this method, diameter distributions for three silica suspensions were obtained from the small-angle X-ray scattering data and also from electron micrographs of the same samples[9]. The agreement between the diameter distributions found by the two methods suggests that small-angle X-ray scattering can be useful for studying diameter distributions in colloidal suspensions of independent spherical particles.

4.2. Outline of the Theory

For a colloidal suspension of independent spherical particles the diameter distribution $\rho(a)$ is given by[8]

$$\rho(a) = (A/a^2) \int_0^\infty dh [\phi(h) - C_4] \alpha(ha) \tag{11}$$

where $\phi(h) = h^4 I(h)$; $h = 4\pi\lambda^{-1} \sin(\theta/2)$; λ is the X-ray wavelength; θ is the scattering angle;

$$C_4 = \lim_{h \to \infty} [h^4 I(h)]$$

$$\alpha(x) = [1 - (8/x^2)](\cos x) - 4[1 - (2/x^2)][(\sin x)/x]$$

and A is a normalizing constant which usually need not be determined but which if necessary can be evaluated by normalizing $\rho(a)$.

Although eq. (11) assumes that $I(h)$ is known for $0 \leqslant h < \infty$, data are actually available only for a finite interval, which will be denoted $h_{min} \leqslant h \leqslant h_{max}$. Techniques therefore are needed for extrapolating $I(h)$ for $h \leqslant h_{min}$ and $h \geqslant h_{max}$ and for estimating the uncertainty resulting from the availability of $I(h)$ data only for a finite interval of h.

Values of $I(h)$ for $0 \leqslant h \leqslant h_{min}$ can be obtained by extrapolating the radius of gyration plot to zero h. This extrapolation will be possible for any scattering curve which is reliable enough to permit meaningful calculation of $\rho(a)$. In this angular region the intensity need be known only with relatively low accuracy, since the factor h^4 in $\phi(h)$ tends to minimize the contribution of $I(h)$ for small h.

Tests indicate that extrapolation of $I(h)$ for $h \geqslant h_{\max}$, though more difficult than for $h \leqslant h_{\min}$, is possible for most scattering curves for which the data extend to h values sufficiently large that $I(h)$ is approximately proportional to h^{-4}. For large h the intensity scattered by polydisperse suspensions has the form[10]

$$I(h) = C_4 h^{-4} + C_6 h^{-6} + C_8 h^{-8} + \ldots \tag{12}$$

where C_4, C_6, and C_8 are constants which can be evaluated by making a few reasonable assumptions about the properties of $\rho(a)$[9]. Under very general conditions[10],

$$\int_0^\infty dh[\phi(h) - C_4] = 0 \tag{13}$$

Also, for almost all suspensions likely to be encountered, all particles can be considered to have diameters greater than some minimum value. Then $\rho(0) = 0$, and[10]

$$0 = \int_0^\infty dh[h^6 I(h) - h^2 C_4 - C_6] \tag{14}$$

From eqs. (12), (13), and (14),

$$(h_{\max})^{-1} \int_0^{h_{\max}} dh\, h^4 I(h) = C_4 - C_6 h_{\max}^{-2} - (C_8/3) h_{\max}^{-4} \tag{15}$$

$$(h_{\max})^{-3} \int_0^{h_{\max}} dh\, h^6 I(h) = C_4/3 + C_6 h_{\max}^{-2} - C_8 h_{\max}^{-4} \tag{16}$$

At $h_{\max}$ the extrapolated intensity given by eq. (12) must equal the measured intensity; therefore,

$$h_{\max}^4 I(h_{\max}) = C_4 + C_6 h_{\max}^{-2} + C_8 h_{\max}^{-4} \tag{17}$$

When the scattering data are used to evaluate the integrals in eqs. (15) and (16), eqs. (15), (16), and (17) form a system of three linear algebraic equations giving the three constants C_4, C_6, and C_8.

With most experimental scattering curves there is no unique way of selecting $h_{\max}$. Best results are not automatically obtained by choosing $h_{\max}$ to correspond to the largest scattering angle at which data can be recorded, since in this angular region the intensity may be so low that the data are not as reliable as data for somewhat smaller scattering angles. A smaller value of $h_{\max}$ therefore may give better results.

Our tests suggest that it is advisable to compute C_4, C_6, and C_8 from eqs. (15), (16), and (17) for each of a series of reasonable values of $h_{\max}$. According to these tests, the most probable form of $\rho(a)$ is obtained for the value of $h_{\max}$ which makes C_8 most nearly equal to zero. This condition is based on the general result[10] that $C_8 = 0$ when $\rho'(0) = 0$, as would be expected under the assumptions with which eq. (14) was obtained.

Figure 4 illustrates this procedure for determining C_4, C_6, and C_8. Further examples are given in Schmidt and Brill[10].

4.3. Results for Silica Suspensions

The procedure outlined in Section 4.2 was applied to Samples I, II, and III, which were silica suspensions in which the particles were spherical.

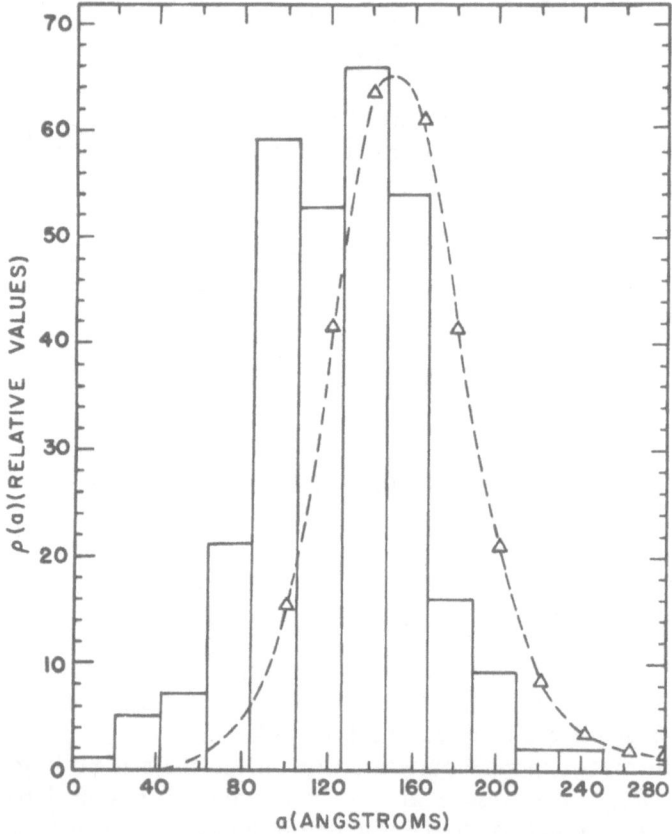

Figure 2. The particle-diameter distribution for Sample I. The histogram gives the results of measurements of particle diameters on electron micrographs, and the triangles and dotted curve for $a \geqslant 100$ Å show the particle-diameter distribution calculated from eq. (11) and multiplied by a constant chosen to permit comparison with the histogram. The portion of the dotted curve for a values less than 100 Å is a reasonable extrapolation of the results computed from eq. (11) for larger a.

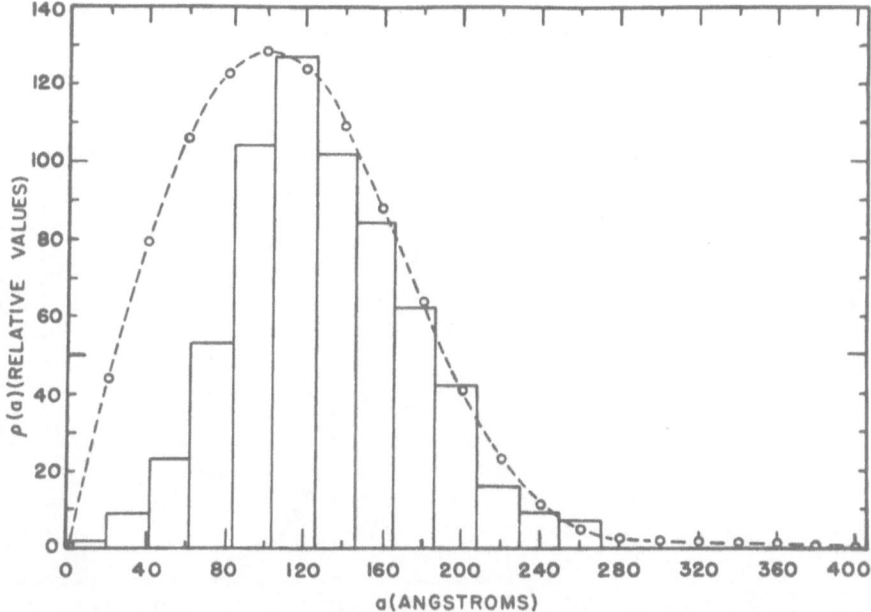

Figure 3. The particle-diameter distribution for Sample II. The histogram shows the results of measurements of particle diameters on electron micrographs. The circles and the dotted curve give the diameter distribution calculated from eq. (11) and multiplied by a constant chosen to permit comparison with the histogram.

Figure 1 shows the scattering curves for these samples. The diameter distributions calculated from these scattering curves are given in figures 2, 3, and 4.

To have an independent check of the reliability of these diameter distribution functions, $\rho(a)$ was also determined for Samples I and II by measuring particle diameters on electron micrographs. The histograms in figures 2 and 3 indicate the diameter distributions obtained from the electron micrographs. The two methods give results which agree within experimental uncertainty.

In Sample I, $\rho(a)$ is narrower than in Sample II, as indicated by the comparison of average particle diameters discussed in Section 3.

Figure 4 shows diameter distributions calculated for Sample III for different choices of h_{max}. According to the criterion of Section 4.2, the curve for $\theta_{max} = 18$ mrad is the most probable diameter distribution, since for this curve the magnitude of C_8 is smallest. In this sample the particles were so small that no attempt was made to find $\rho(a)$ from electron micrographs. Probably only the X-ray method can give diameter distributions in samples made up of such small particles.

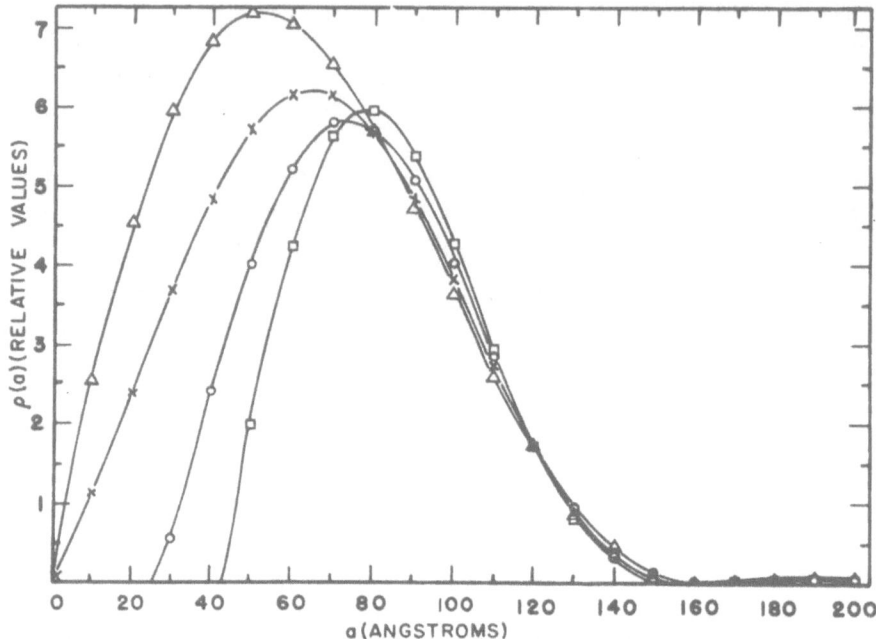

Figure 4. The particle-diameter distribution $\rho(a)$ for Sample III. For the curves shown by triangles, crosses, circles, and squares, h_{max} corresponded to scattering angles of 17, 18, 19, and 21 mrad, respectively. For each value of h_{max}, the constants C_4, C_6, and C_8 were computed from eqs. (15), (16), and (17). The reliability criterion suggests that the curve shown by crosses is the most probable particle-diameter distribution.

4.4. Discussion

It is difficult to make a quantitative estimate of the magnitude of the errors which can be expected in values of $\rho(a)$ calculated from the scattering data. The uncertainty definitely exceeds the uncertainty present in the original scattering data. Nevertheless, according to our experience with the theoretical test functions and the silica samples, diameter distributions calculated from eq. (11) can be expected to be at least as reliable as the results obtained from electron micrographs of such small particles.

Anomalous results are sometimes obtained for $\rho(a)$ for diameters smaller than the value at which $\rho(a)$ has its principal maximum. For example, sometimes the calculations can yield negative values of $\rho(a)$. Such results must be artifacts, since $\rho(a)$ is defined to be a probability density function, for which negative values have no meaning. The value of h_{max} can be chosen to minimize these effects.

Negative values for small a were obtained for the $\rho(a)$ curves for Sample III when h_{max} corresponded to angles of 19 and 21 mrad. [These values of $\rho(a)$ were not plotted in figure 4.] Also, the results obtained from eq. (11) for $\rho(a)$ for Sample I were obviously unreasonable and therefore replaced by the extrapolated curve shown in figure 2. The occurrence of such obviously incorrect values is not a serious problem since they are found only for relatively small values of a, where they can easily be seen to be spurious and where the form of $\rho(a)$ is often of little interest.

The choice of the extrapolation constants C_4, C_6, and C_8 has little effect on most of the $\rho(a)$ curve. For example, in figure 4 the form of $\rho(a)$ in the neighborhood of the maximum and in the outer part of the curve is seen to be nearly independent of the value selected for h_{max}.

Because of the difficulty of obtaining any information about diameter distributions in suspensions of such small particles, diameter distributions calculated from the X-ray data may be of considerable value in spite of the relatively large uncertainty. When $\rho(a)$ is used only to compute average quantities, it need be known with less accuracy than in cases where $\rho(a)$ itself is of primary interest.

Computer programs have been prepared for making the necessary calculations, and the evaluation of $\rho(a)$ is suggested as a regular procedure in the analysis of the small-angle X-ray scattering data from polydisperse suspensions of spherical particles. Further information about the computer programs is available from the authors.

ACKNOWLEDGMENTS

The authors express their appreciation to James E. Thomas and T. Bruce Daniel for assistance and advice during the investigation, to Ted R. Taylor for help in recording the scattering data from the silica samples, to Gordon Shaw and his associates at the Midwest Research Institute for obtaining the electron micrographs, and to the staffs of the computing centers of the University of Missouri at Columbia and Kansas State College at Pittsburg for aid with the numerical calculations.

The authors are very grateful to J. Kratohvil of Clarkson College and to J. Bugosh, R. L. Rusher, and co-workers of the du Pont laboratories for supplying the silica samples and for advice and information concerning the use and properties of these samples.

REFERENCES

1. A. Guinier, G. Fournet, C. B. Walker, and K. L. Yudowitch, *in: Small-Angle Scattering of X-rays*, John Wiley, New York (1955).
2. P. W. Schmidt, *J. Phys. Chem.* **69**, 3850 (1965).
3. P. W. Schmidt and R. Hight, Jr., *J. Appl. Phys.* **30**, 867 (1959).
4. P. W. Schmidt, unpublished research (1967).
5. P. Mittelbach, *Kolloid-Z.* **206**, 152–159 (1965).
6. L. C. Roess, *J. Chem. Phys.* **14**, 695–697 (1946).
7. J. Riseman, *Acta Cryst.* **5**, 193–196 (1952).

8. J. H. Letcher and P. W. Schmidt, *J. Appl. Phys.* **37**, 649–655 (1966).
9. L. Brill, C. G. Weil, and P. W. Schmidt, research to be published. An extended account of part of this investigation is contained in the thesis presented by O. L. Brill in partial fulfillment of the requirements for the PhD degree at the University of Missouri (1967). (Copies available from University Microfilms, Ann Arbor, Michigan.)
10. P. W. Schmidt and O. L. Brill, *in Proceedings of the Second Interdisciplinary Conference on Electromagnetic Scattering* (Amherst, Massachusetts, 1965) (R. L. Rowell and R. S. Stein, eds.), Gordon and Breach, New York (1967), pp. 169–186.

VI. SMALL-ANGLE X-RAY SCATTERING AND LOW-ENERGY ELECTRON DIFFRACTION STUDIES OF CATALYST SURFACES

Gabor A. Somorjai

Inorganic Materials Research Division, Lawrence Radiation Laboratory
Department of Chemistry, University of California
Berkeley, California

Small-angle X-ray scattering can be conveniently used to study the proper-ties of supported catalyst surfaces. The size distribution and growth kinetics of platinum particles (10 to 300 Å) in an alumina matrix were measured as a function of temperature in an oxidizing and reducing atmosphere. The scattering background due to the porous alumina has been minimized by pressing the samples at 100 kbar prior to the measurements. The mecha-nism of the particle growth has been determined.

The (100), (110), and (111) surfaces of platinum single crystals have been studied by low-energy electron diffraction. The presence of surface structures, ordered and disordered, which exist in different temperature ranges have been uncovered. The surface Debye temperature of the different crystal faces have been measured. The properties of the surface structures will be discussed.

1. INTRODUCTION

Studies of solid surfaces and chemical reactions which occur at the interface are among the most important and exciting fields of chemistry. The abun-dance and variety of surface catalytic processes indicate the diversity of roles the solid surface seems to play during the reaction. The largest obstacle, however, which prevents closer scrutiny of the structure of sur-faces and the detailed mechanism of a surface reaction, has been the scarcity of experimental tools which permit one to study surface properties on an atomic scale.

Low-energy electron diffraction (LEED) proves to be a novel experi-mental technique which permits the study of the structure and rearrange-ments of atoms in the surface plane. Since the electrons do not penetrate below a few atomic planes (four–five) the diffraction patterns are representative of the surface arrangement of atoms. Thus, this technique is sensitive only to surface properties. Low-energy electron diffraction plays the same role in analyzing the structure of solid surfaces as X-ray diffraction in studies of the bulk structure. The technique can also be employed for studies of the surface arrangement of adsorbed gases and for investigations of chemical surface reactions.

There is also widespread use in catalytic surface reactions of polycrystalline metal particles of size 10 to 10^3 Å which are dispersed on highly porous supports. Typical supported catalysts include platinum, palladium, or nickel supported on alumina or silica. There has been a lack of effective experimental techniques which allow studies of the sizes and shapes of such particles and the changes that occur during chemical reactions. Changes in particle size, i.e., in the surface-to-volume ratio, can markedly influence the rates of catalytic surface reactions. Small-angle X-ray scattering proves to be the most versatile technique to detect and monitor particles ranging from 40 to 600 Å. The only criterion for monitoring the changes which affect the particle shape or size in this range is the presence of an electron density difference between the metal particles to be investigated and the support matrix, since the scattered intensity is proportional to the square of the difference in electron density[1].

In this paper we shall discuss two studies of platinum surfaces: (1) low-energy electron diffraction study of the clean (100), (111), and (110) faces of platinum single crystals and (2) small-angle X-ray scattering investigation of the effect of thermal history on the particle size of platinum catalysts dispersed on highly porous η-alumina.

These studies should give an indication of the types of experimental information which could be obtained by these techniques.

2. PRINCIPLES OF LOW-ENERGY ELECTRON DIFFRACTION

In low-energy electron diffraction studies one uses monochromatic electrons of energies $E \simeq 5 - 500 \pm 0.2$eV, which correspond to a wavelength range of roughly $0.5 - 5$ Å, since $\lambda(\text{Å}) = [150/E(\text{eV})]^{1/2}$. The most versatile low-energy electron diffraction apparatus for catalytic studies seems to be the post acceleration type (Varian Associates, Palo Alto, California). The schematic representation of the experiment is shown in figure 1. The electrostatically focused electrons ($\approx 1~\mu$A) are back reflected from the crystal surface due to their charge and large mass. The crystal and the first grid are at ground potential to assure a field-free region in order to minimize distortion of the path of the scattered electrons. The second grid serves to retard the inelastically scattered fraction of electrons,

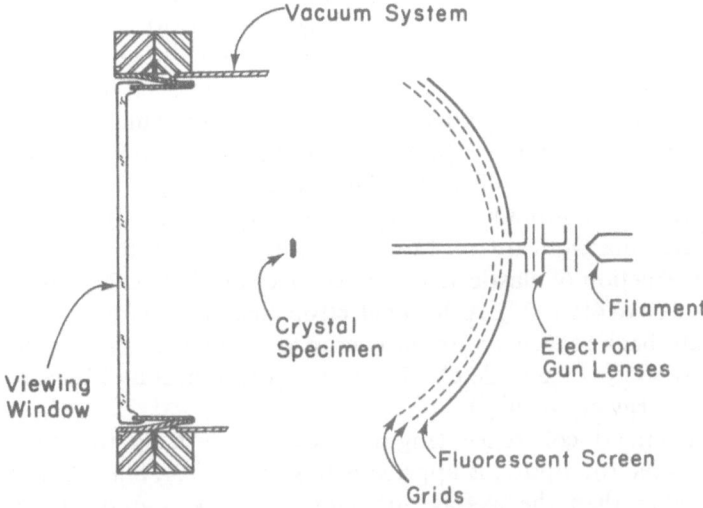

Figure 1. Scheme of the low-energy electron diffraction experiment.

and it is held generally at cathode potential. The elastically scattered electrons, which contain the diffraction information, penetrate the second grid, are accelerated, and strike a fluorescent screen where the diffraction pattern which corresponds to the surface structure is displayed. Although better focusing([2]) and energy selection([3]) techniques have been reported which could give almost an order of magnitude better resolution with some sacrifice of the beam intensity, the electron optics which are described above proved to be entirely adequate for catalytic surface reaction studies.

Our studies on platinum surfaces indicate that in the range of 5 to 50 eV the diffraction is essentially two-dimensional. Above this energy two to five atomic planes seem to contribute to the diffraction. The surface structures which are the property of the topmost layer of atoms are visible in the energy range of 5 to 150 eV. At higher energies (> 150 eV) electrons penetrate deeper below the surface and the relative contribution of atoms in the surface plane to the total scattered intensity diminishes.

The scattering amplitude of electrons by atoms is approximately three orders of magnitude greater than that of X-rays([4]). This implies a six orders of magnitude greater scattered intensity. Thus, electrons are such efficient scatterers that only a small concentration of surface atoms which are ordered in regular arrays are needed to give detectable diffraction patterns. Patterns which are due to atomic arrangements occupying less than 5 % of the total number of surface sites ($\sim 5\%$ coverage) have been detected([5]). The surface domains which give rise to the diffraction patterns contain on the average 40 atoms in ordered arrays[6]. The observed intensities come from the coherent superposition of electron waves scattered by these separate domains.

Due to the large scattered intensity there is no need for detection techniques which are accumulative in a long time span. Thus, visual or photographic detection of the diffraction patterns which are displayed on the fluorescent screen is sufficient. The intensity changes of the diffraction spots can be monitored by a high sensitivity, small-angle spectrophotometer (Gamma Optical Co., Model 2000, was used in these studies).

The fraction of elastically scattered electrons is 5 to 10 % in the energy range of 100 to 450 eV[5]. At lower electron energies the fraction of electrons which are back-reflected without energy loss increases sharply and attains the 50 to 80 % range at 20 eV. Thus, the surface becomes highly reflective for low-energy electrons[5].

The lateral coherence length of electrons, using the commercially available electron optics, is approximately 15 to 20 Å. Since this is appreciably smaller than the average domain size ($\sim$40 atoms $\sim$120 Å), direct detection of atomic steps is difficult[6].

In low-energy electron diffraction studies one face of a high purity single crystal of size appreciably larger than that of the size of the electron beam (1 mm^2) ideally should be used. The experiment is carried out in a vacuum of better then 10^{-8} torr in order to avoid surface contamination due to adsorbed gases. A crystal surface will adsorb a monolayer of foreign gas at 10^{-6} torr in about 1 sec (assuming a sticking coefficient of unity), so that typical operating pressures in the diffraction chamber are 10^{-8} to 10^{-10} torr to allow reasonable observation times of the clean surface. The diffraction apparatus and the high vacuum system which are necessary for these surface studies are available commercially at a reasonable cost.

The use of a mass spectrometer which can be attached to the diffraction chamber is necessary for studies of chemical surface reactions. The availability of quadrupole mass spectrometers in recent years (Electronic Associates Model 210 was used in our studies), which are compact and have high detection sensitivity (10^{-12} torr partial pressure), have aided greatly the definitive interpretation of the experimental results. The mass spectrometer allows one to monitor changes in the mass spectrum which are caused by less than 1 % of the monolayer provided that the desorption takes place in short times (about 1 min). The use of a mass spectrometer can also greatly aid the detection of surface contaminants.

The single crystal surface area on which the studied chemical surface reaction takes place is much smaller than the wall area of the diffraction chamber. Therefore, care should be taken that the reactants and the reaction products do not undergo chemical changes when colliding with the wall of the chamber. If the reaction which is to be studied is influenced by collisions with the chamber wall the experiment could be designed to assure that the reactant gas species impinge directly onto the surface of the single crystal and that the product species can enter the ionizer region of the mass spectrometer without colliding with the chamber walls first.

Low-energy electron diffraction studies of solid single-crystal surfaces have revealed a great deal of information about the physical–chemical properties of surfaces which could not have been obtained by other experimental techniques[5]. One of the striking results of low-energy electron diffraction studies on clean solid surfaces is the discovery that the substrate atoms may reside in surface structures of different kinds[7]. The presence of these surface structures is indicated by the appearance of extra diffraction features, which are superimposed on the diffraction pattern of the substrate unit mesh predicted by the bulk unit cell. One can assign lattice parameters to these surface structures which, in general, were found to be integral multiples of the lattice parameters which characterize the substrate. Studies on clean semiconductor and metal surfaces have revealed the presence of several surface structures with unit cells in the range of twice (2×2) or as large as 8 times (8×8) the unit cell dimensions of the substrate[5,8]. The exact arrangement of atoms which produce these structures is still in question. Their appearance, however, indicates (1) long-range order on the surface and (2) that transformation (reversible or irreversible) from one type of surface structure to another can readily occur. Thus, these structures which may be called surface phases have a temperature range of stability, and phase transformations at the surface can take place without any apparent effect on the bulk structure.

There are several reasons for the existence of surface structures which are indicated by recent experiments[5]. The activation energy for surface diffusion of adatoms is appreciably smaller than either the activation energy for bulk diffusion or heat of sublimation in monatomic solids[9]. Thus, surface atoms are shielded by large potential-energy barriers from rapid exchange with the bulk or vacuum. The phonon spectrum of surface atoms is different from that of the bulk atoms as shown by surface Debye temperature measurements[10,11]. There is also some experimental evidence that, in addition to the change in the mean-square displacement of atom at the surface, there is a net expansion or contraction with respect to the bulk lattice spacing[12,13].

Another important discovery of low-energy electron diffraction studies is the observation that adsorbed gases form a variety of ordered surface structures on the different solid surfaces[5]. The type of structure which forms depends on the crystallographic orientation and temperature of the substrate and the size, surface concentration, and chemical nature of the adsorbed species. Structural transformations easily occur as a function of changes in these parameters. Consequently, a great wealth of surface structures which can be attributed to the adsorbed species have already been reported([5]).

These experimental results seem to indicate that a study of surface structures which are due to the solid atoms or to adsorbed gas atoms is essential to understand the nature of catalytic surface reactions. It is already apparent that the diversity of surface catalytic properties can be correlated with the variety of surface phases and structures which seem to form as a function of experimental conditions. Low-energy electron diffraction can provide us with such structural information.

3. LOW-ENERGY ELECTRON DIFFRACTION STUDY OF THE (100), (111), AND (110) FACES OF PLATINUM

3.1. Experimental

Platinum single crystals of the highest purity (99.9999 %) were used in the experiments. The samples (0.5 to 2 mm thick and 6 mm in diameter) were cut after orienting the particular face by X-ray within 1°. The crystals were then polished, etched; and spot welded to a platinum holder. After spotwelding thermocouples to the back of the crystal, the sample was introduced into the diffraction chamber. Baking and degassing the evacuated diffraction chamber produced pressures of the order of 5×10^{-10} torr.

Ion bombardment using high purity xenon or argon was employed in order to (1) remove the surface damage which was caused by the mechanical surface treatments or (2) remove disordered surface structure which formed irreversibly at high temperatures ($T > 750°C$). Usual conditions of ion sputtering were 2×10^{-5} torr A or Xe, 340 eV accelerating potential for 2 hr. Heating the samples to $T > 750°C$ has also produced diffraction features without ion bombardment. All of the surface structures reported in this paper were reproducible under all conditions of surface preparations and ion bombardment treatments unless otherwise noted.

In order to monitor the composition of the ambient and detect the presence of possible surface contaminants, a quadrupole mass spectrometer (EAI 210) was permanently mounted on the diffraction chamber. This way the background ambient composition during heat treatments and the composition of the gases used for ion bombardment could be continuously monitored.

In order to further probe the nature of the different surface structures, they were heated in oxygen and in hydrogen in their temperature range of stability. In this manner, surface contaminants which form volatile oxide products or react in a reducing atmosphere could be detected by the mass spectrometer and eliminated from the surface.

The diffraction patterns yield a great deal of information about the structures of the surface. Intensity analysis of the diffraction features is necessary, however, to distinguish between several structures which could yield similar diffraction patterns[5]. This analysis is carried out by monitoring the intensity of the diffracted beam as a function of electron energy and scattering angle. Variation of the intensity of a given diffraction spot with heat treatment or in the presence of gases can give information on the kinetics of surface diffusion or adsorption. Intensity measurements were made using a telephotometer with fiber optics (Gamma Scientific Instruments, Model 2000) which allowed highly reproducible measurements of intensity fluctuation with beam voltage which were plotted on an x-y recorder. These were then taken at various angles of incidence. The fluorescent screen has also been photographed directly to obtain the diffraction patterns. The relative intensities could then be obtained by using a densitometer.

Pattern a: Diffraction pattern of the clean Pt(100) face.

3.2. Surface Phase Transformations as a Function of Temperature

Annealing the crystals at 600°C for 1 hr after the ion bombardment has produced the diffraction patterns which are predicted by the bulk unit cell. Using the notation that is suggested by Wood, this corresponds to the (1 × 1) or substrate structure (pattern *a*).

If the samples are annealed after ion bombardment in other temperature ranges, either below or above 600°C, new diffraction spots appear which indicate that the atoms at the surface are arranged in surface structures which are different from those predicted by the bulk unit cell. These surface structures can be easily reproduced on all of the single crystal sample used. All of the structures which were observed on the different platinum substrates are summarized in table I. The structures are divided into two types, ordered and disordered. The ordered surface structures are stable only at temperatures roughly below one half the melting temperature and characterized by unit cells which are integral multiples of the substrate unit cell. The diffraction pattern of one of these surface structures, the (5 × 1) structure, is shown in pattern *b*. The disordered structures appear at high temperatures. All of the experimental observations seem to indicate that both types of surface structures are the property of the clean platinum surfaces. No single impurity produces the variety of diffraction features which were observed as a function of substrate temperature in the high

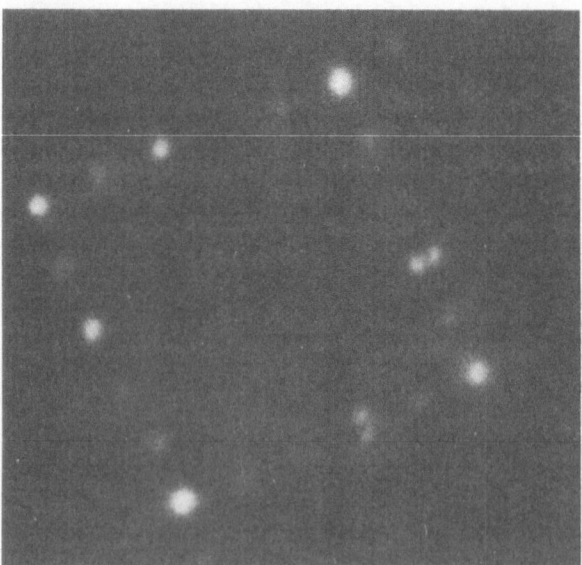

Pattern *b*: Diffraction pattern of the (5 × 1) surface structure on the Pt(100) face.

Table I. List of Surface Structures Which Were Detected on the Different Low-Index Surfaces of Platinum and Their Approximate Temperature Range of Stability

Substrate	Surface structure	Approximate temperature range of stability
Pt (100)	(5 × 1)	350–500°C
	(2 × 1)	300–500°C
	⊙[a]	>700°C
Pt (100)	⊙[a]	>600°C
Pt (111)	(2 × 2)	800–1000°C
	(3 × 3)	800–1000°C
	⊙[a]	>900°C

[a] The notation ⊙ indicates a ringlike diffraction pattern.

purity platinum surfaces. The surface structures show a broad range of different physical properties (stability range, long- or short-range order, and different types of surface structures on each substrate). Heat treatment in an ultrahigh vacuum and in oxidizing or reducing ambients has not produced any new volatile species detectable by the mass spectrometer.

3.2.1. Properties of the Ordered Surface Structures

The following statements summarize the experimental information concerning the ordered surface structures which appear on the different platinum substrates.

1. The surface structures which appear on the different crystal faces of platinum are stable only in well-defined temperature ranges. The stability range of two structures may overlap on the given substrate.

2. The surface structures are not affected by heat treatments in oxygen or hydrogen in their temperature range of stability.

3. The structures are in registry with the substrate (not rotated) and are characterized by lattice parameters which are integral multiples of that of the substrate [(5 × 1), (2 × 1), (2 × 2), and (3 × 3)].

4. The surface structures anneal out at temperatures below their range of stability but reappear readily when reheated in their stability range. Once heated above this temperature range they disappear irreversibly.

5. The surface structures were obtained only after ion bombardment. The properties of the surface structures, however, were independent of the type of ions which were used in the ion bombardment.

6. The surface structures could rapidly be obtained by applying a steep temperature gradient along the gradient surface. The structures could also be formed on the (100) substrate by heating the crystals in a well-defined temperature range.

7. The intensity of the diffracted beams emanating from the surface structures is of the same order of magnitude as the intensity of the substrate reflections.

8. The different heat treatments have not resulted in any appreciable rise in the ambient pressure or in the appearance of volatile impurities, either when the heating was commenced or during the heating period.

9. The presence of these surface structures causes no change in the positions of the intensity maxima in the I_{00} vs. E (eV) curves which were taken for the clean substrates.

10. Several surface structures were found to exist on the (100)-face of silver, gold, and palladium single crystals[14,15]. The ordered surface structures on the gold substrate are similar to those found on the (100)-face of platinum.

3.2.2. *Properties of the Disordered Surface Structures*

When the (100) substrate is heated above 700°C a new diffraction pattern slowly appears. This is characterized by narrow, circular segments (pattern c). Six such segments appear at first, then with increased heating time or high heating temperature, 12 and then 24 segments form which finally join into a ring (pattern d).

The ringlike patterns can be obtained without ion bombardment only by heating the sample between 700°C and the melting temperature (1769°C) Their formation is irreversible, i.e., once they have been created they are stable indefinitely at any temperature below the melting point and can be removed only by ion bombardment.

If the sample is heated above 1000°C for an extended period (4 to 6 hr), the intensity of the diffraction features which are due to the substrate unit mesh and which coexist with the ring pattern decreases while the intensity of the rings increases. When heated near the melting point, the ringlike diffraction pattern finally remains the only diffraction feature on a presumably disordered surface.

The rings are concentric about the (00)-reflection, have sharp outlines, and show the usual diffraction features of other surface structures; (1) they appear only at low electron energies (5 to 150 eV) and (2) they appear at decreasing angles with respect to the (00)-spot with increasing electron energy. One striking feature of the ring pattern is that it does not overlap with any of the diffraction spots which are due to the substrate unit mesh.

Pattern *c*

Pattern *d*

Patterns *c* and *d*: Development of the ring-diffraction pattern on the Pt(100) surface as a function of heating time or increased heating temperature.

Ringlike diffraction patterns were found to exist on both the (111) and (110) platinum surfaces as well.

The following statements summarize some of the important properties of the disordered surface structures and other pertinent information which could be used to interpret these structures:

1. Ringlike diffraction patterns form on the (111), (110), and (100) substrates of platinum via sets of radially symmetric segments at elevated temperatures.

2. Their formation is irreversible, and they become the only diffraction feature of the surface as the melting point is approached. They may be formed without the use of ion bombardment by heating the substrate in their temperature range of stability.

3. The ring patterns are unaffected by heat treatment in hydrogen or oxygen atmospheres.

4. The rings appear to have apparent lattice parameters which are smaller than the smallest interplanar distance in the substrate plane.

5. The surface structure which gives rise to the ring pattern is parallel to the surface upon which it forms and shows diffraction features similar to other surface structures.

6. The ringlike patterns are narrow and well defined and always appear in the same position on a given substrate. They are unlike the radial distribution functions observed by neutron diffraction[16] or X-ray diffraction studies of liquids near the melting point[17].

7. The formation of these ring patterns reflects the gradual loss of long-range order at the surface as heating time or temperature is increased.

8. Ringlike diffraction patterns were also found to form on other metal surfaces[14,15]. They have apparent lattice dimensions which are smaller (Ir, Au) or larger (Ag) than the distance of closest approach in the ordered substrate.

9. There is no evidence of any macroscopic change in the surface structure of platinum as observed in the photomicrograph (1000 × magnification) when the ringlike diffraction pattern is formed.

The lattice parameters and the same ratio of lattice spacings which can be assigned[18] to the different rings ($d_O^I/d_O^{II} = \sqrt{3}$ and $d_O^I/d_O^{III} = 2$) on all three substrates suggest that the ringlike diffraction patterns are due to domains of (111) surface structures on all faces of platinum with reduced nearest-neighbor spacing. These hexagonal surface structures appear at preferred orientations at first, as shown by the presence of ring segments. After extended heating time or as the melting temperature is approached they can be freely rotated in the substrate plane. The disordered hexagonal surface structures show an 11% contraction with respect to the interplanar spacings in the ordered (111) face.

In addition to the ringlike patterns on the platinum surfaces, ring or segmented ringlike surface structures have been observed on gold (100), silver (100), and iridium (111) surfaces[14,15]. In no case do the diffraction rings coincide with the diffraction spots of the substrate, but they show "contraction" or "expansion" of different magnitudes. For example, in silver the apparent lattice parameter which can be assigned to the ring pattern indicates a 13% expansion with respect to the nearest-neighbor distance in the ordered substrate while gold shows a "contraction" which is similar to that of platinum. It is likely that the disordered phase is present at high temperatures on many face-centered cubic metal surfaces which have low enough evaporation rates not to permit the removal of the new phase into the vapor as soon as it forms.

4. PRINCIPLES OF SMALL-ANGLE X-RAY SCATTERING

The small-angle X-ray measurements were made using Cu $K\alpha$ radiation, a four-slit collimating system, and a sample-detector distance of 50 cm. Figure 2 shows a schematic representation of the instrumental arrangement. Although this apparatus was built in our laboratory, several small-angle X-ray scattering attachments are available commercially which assures easy applicability of this technique. The sample was mounted on a 1-mil Mylar sheet behind the third slit. The X-ray intensity was measured using scintillation counting with pulse-height selection[19,20].

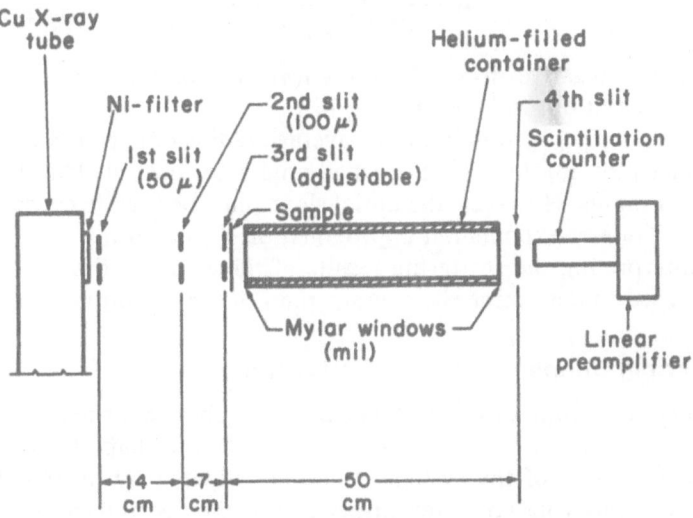

Figure 2. Scheme of the small-angle X-ray scattering experiment.

The smallest angle which did not strike the primary beam was 0.0733°, or about 4.4 min of arc. The largest angle at which these experiments gave scattering intensity distinguishable from the background was about 1.1°. Roughly, these angles correspond to particles ranging from 40 to 600 Å.

The pulses leaving the analyzer (10) were counted with a conventional scale-of-256 and a mechanical register. The typical counting rate was some 2000 pulses min, although it was much higher near the primary beam and much lower at angles near 1°.

From the small-angle X-ray scattering, the datum that can be calculated most directly is the average radius of gyration of the scattering particles. For a detailed discussion of the theory, the reader is referred to Guinier's book([1]).

The scattering intensity $I(h)$ for N = identical noninteracting particles is given as

$$I(h) = I_e N \overline{F^2(h)} \qquad (1)$$

where I_e is the intensity of scattering by one electron; $F(h)$ is the structure factor; and $h = (4\pi/\lambda) \sin \theta$, where λ is the X-ray wavelength and 2θ is the scattering angle. At small angles $\overline{F^2(h)}$ can be approximated by $\overline{F^2(h)} = n_e{}^2 \exp(-h^2 R^2/3)$, where R is the radius of gyration and n_e is the total number of electrons in the particle. For small angles $[(4\pi/\lambda) \sin \theta \simeq (4\pi/\lambda)\theta]$, R is given by

$$R = \frac{1}{2\pi}\sqrt{\frac{3}{\log_{10} e}}\lambda\sqrt{p} \qquad (2)$$

where p is the negative slope of the log $I(h)$ vs. $(\tan 2\theta)^2$ curve. In case of the Cu $K\alpha$ radiation, $R(\text{Å}) = 0.645 \sqrt{p}$.

The usual plot for obtaining the average radius of gyration is therefore the Guinier plot, log I vs. h^2. For small angles $h \sim 2\pi\lambda^{-1}\theta$. The slope gives the desired radius. However, the initial slope must be used because at larger angles the Guinier exponential approximation begins to fail.

In interpreting the scattering results of this investigation, the Guinier plot was used to determine the average radii of the platinum particles.

4.1. Particle Shape and Size Distribution

The Guinier approximation holds best for spherical or nearly spherical particles. However, there is no way to determine the shape from R alone. Furthermore, most of the systems to be investigated by means of small-angle X-ray scattering are heterodisperse, and the distribution of particles is of great interest. Shull and Roess([21]) developed a method for calculating small-angle scattering intensities of heterodisperse systems, inserting a distribution function into the scattering integral. It was assumed that all particles are geometrically similar; i.e., the number of electrons n_e is proportional to R^3, where R is the radius of gyration.

Then

$$I(h) = K \int_0^\infty \overline{F^2(h)} N(R) R^6 \, dR \qquad (3)$$

where $N(R)$ is the number distribution function and $N(R) \, dR$ represents the total number of particles in the size range from R to $R + dR$, and K is a constant. In our work on platinum catalysts, the exact structure factor for spheres was used instead of the Guinier approximation. The calculated particle sizes agreed within 10% with those obtained by using the Guinier approximation. In case $N(R)$ is a Maxwellian distribution,

$$N(R) = R^n \exp - (R/R_0)^2$$

where R_0 and n are constants.

If one writes for a spheroid that $V(r_e) = (4\pi/3) v r_e{}^3$, where v is the axial ratio and r_e is the equatorial radius, and substitutes into the scattering intensity integral, this integral can be readily evaluated at three different limits. For spherical particles $v = 1$, for disc-shaped particles $v \to 0$, and for rodlike particles $v \to \infty$ and $r_e \to 0$ simultaneously, so the product rv approaches L, the length of a rod with negligibly small radius. Expanding the integral by hypergeometric functions—see Roess and Shull[21]—one gets the scattering from a distribution of particles by different shapes. These properties of the scattering from polydisperse samples are discussed in the chapter by Paul M. Schmidt, which deals with small-angle scattering methods for determining the distribution function $N(R)$.

By comparison of the experimental data with the theoretical scattering curves, one should be able to get a distribution of the scattering particles and find some information about the general shape of the particles. There are various ways of making graphical comparisons of the data with the theories, one convenient type of plot being that $\log I\theta^2$ vs. $\log \theta^2$, which gives curves with distinct maxima on which the effects of shape and distribution function are fairly readily discernible. In figure 3 these curves are plotted for monodisperse spheres and for various Maxwellian distributions. Analogous curves for discs and for rods are also available [19,21].

The experimental scattering data for platinum on alumina were examined in the light of these theoretical curves, and such conclusions as could be drawn will be taken up with the subsequent discussion of the experimental results.

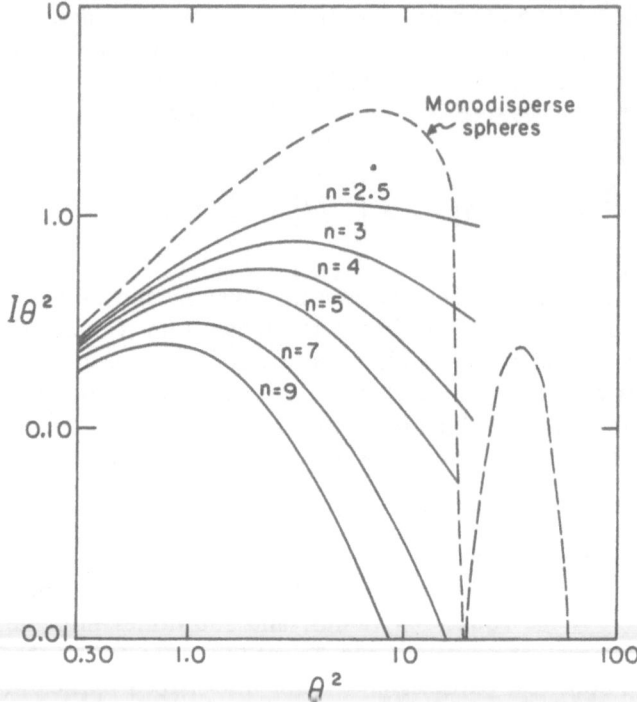

Figure 3. Scattering curves for monodisperse spheres and Maxwellian distribution of heterodisperse spheres—monodisperse spheres.

5. SMALL-ANGLE X-RAY SCATTERING STUDY OF ALUMINA-SUPPORTED PLATINUM CATALYSTS

There is a major practical difficulty in the small-angle investigation of microporous catalysts; the intense scattering arising from the holes in the catalyst support itself. Consider, e.g., the metallized catalyst used in the present investigation, platinum supported on η-alumina. The electron density of platinum is $78 \times 21.45/195.09 = 8.58$ faradays/cm^3, that of the crystallites of η-alumina approximately $30 \times 3.7/60 = 1.85$ faradays/cm^3; and of the holes very nearly zero. Since the absolute intensity of the X-ray scattering is proportional to the square of the difference in electron density, a given volume of holes will scatter about one-thirteenth the intensity of the same volume of platinum. The scattering from holes is easily large enough to permit the small-angle investigation of the particle sizes in the microporous solids themselves, and there are a number of examples in the literature of studies on finely divided solids, including catalysts([1]). Even much smaller differences in electron density are sufficient for small-angle scattering([5]): 0.05 faraday/cm^3.

If the location of the holes were random with respect to the platinum particles, the scattering from the holes could be treated as part of the background correction and subtracted out. But the platinum particles are sitting within the holes, so that their scattering tends to cancel one another. In fact, our preliminary experiments showed that the scattering from platinum–alumina was apparently less than that from the alumina itself. Therefore, it is essential to destroy the holes before studying the scattering from the platinum.

A possible method would be to fill the holes with a liquid of the same electron density as the alumina; 1.85 faradays/cm^3. It has been shown([23]) that the hole scattering from a silica–alumina cracking catalyst of electron density 1.245 was diminished about twofold by saturating it with o-xylene and diminished more than a hundredfold by saturating it with n-butyl iodide. We have found a better and simpler method to eliminate scattering from the holes.

We have found that by pressing the supported catalysts samples in a hydrostatic press([24]) under a pressure of 100 kbar or more, substantially all holes in the alumina are reduced to a size giving no small-angle scattering; the gases originally present escape through the material of the press. In the "pressure sintered" sample the scattering by the metal particles is readily measured.

With the development of pressure sintering as a method for the elimination of hole scattering, it becomes possible to study the effect of such variables as the method of original impregnation, the thermal history of the metallized catalyst, or the chemical attack upon the metal particles. In the present investigation the second of these problems was studied: the effect of thermal history on particle size. The catalyst chosen for investigation was platinum on alumina. It has been possible to follow the growth of the platinum particles and to derive information about the kinetics and mechanism of the process.

5.1. Experimental

The catalysts subjected to study were 5 and 0.5% platinum by weight, dispersed on high-area microporous alumina. The alumina is characterized by its X-ray diffraction pattern as η-alumina; its crystal structure is not affected by heating to the temperatures used in this investigation; and X-ray diffraction patterns taken on our samples after pressure sintering showed no change in its crystal structure even after compression to 350,000 atm. The catalyst pellets were ground and screened through a

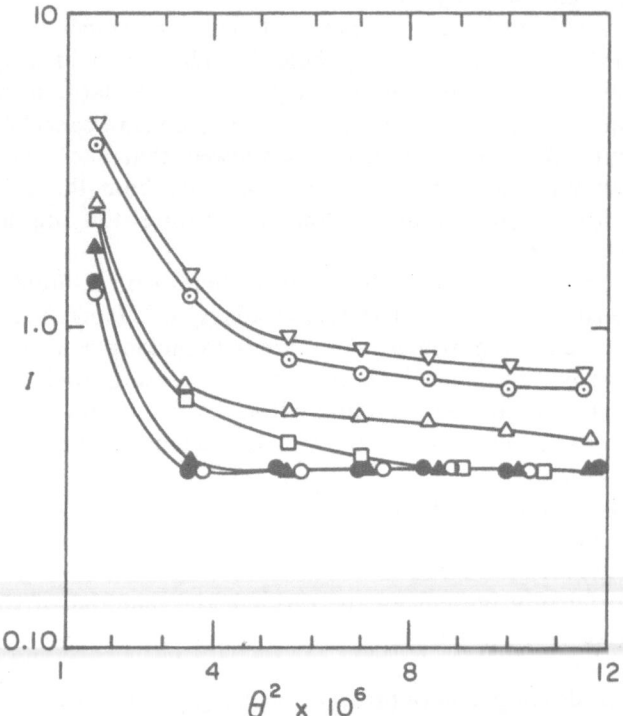

Figure 4. Scattering intensity of unheated η-alumina at various pressures. i intensity in units of 256 counts/30 sec. θ angle in radians. ● air background alone, no sample; ▲ alumina pressed at 100,000 atm; ○ alumina pressed at 300,000 atm; □ alumina pressed at 65,000 atm; △ alumina pressed at 40,000 atm; ◉ alumina pressed at 20,000 atm; and ▽ alumina unpressed.

200 mesh/in. screen. The samples were heat treated (at 400 to 700°C) for the desired time (1 to 96 hr) in an oxidizing atmosphere (air) or a reducing atmosphere (generally illuminating gas, but identical results were obtained with hydrogen).

Every time a platinum–alumina sample was heated, a sample of the pure alumina was heated side by side in the furnace, so that a correction might be made in the X-ray measurement for any changes taking place in the "blank" alumina upon heating. In fact, this precaution proved to be unnecessary since all the alumina samples turned out to give identical

X-ray scattering. After heat treatment, the platinum–alumina samples were compressed at 100,000 atm for 15 min. The detailed description of the high-pressure press is published elsewhere[24].

Figure 4 shows the X-ray scattering intensity of unheated pure alumina treated at various pressures. As it shows, the scattering from holes in the alumina has been substantially eliminated at 100,000 atm, so this pressure was chosen for the treatment of all samples. Heating in the range of temperatures used has a negligible effect on the compressibility characteristics of the alumina.

There was no change in the shape of the platinum particles due to compression in the pressure range of 5 to 200 kbar, as was determined from the small-angle X-ray scattering data.

At one stage of the investigation an attempt was made to pressure sinter the platinum–alumina first and then to treat it. The results were quite disconcerting, and the attempt was abandoned: The scattering in the small-angle region increase enormously, producing a background scattering so high that the scattering from the platinum particles could not be determined with any confidence. The cause of this phenomenon is the high strain to which the alumina is subjected upon compression, which greatly facilitates crystallization. It is probable that the grains rearrange to an order such as they had before compression, thereby introducing a number of holes.

5.2. Results and Discussion

5.2.1. *Growth of Platinum Particles*

The body of our platinum–alumina studies were made with 5% platinum catalysts, although 0.5% platinum is also easily detectable by its scattering[19]. The samples were compressed at 100,000 atm after the heat treatments.

Figure 5 shows the particle size changes in a reducing atmosphere at two temperatures, 600°C and 700°C, and figure 6 shows the particle size changes for oxidizing atmospheric heating.

The effect of heat treatment on the particle size is clearly indicated in these data. There is a large increase in the average particle size at these temperatures in the measured time interval. All samples exhibit a very fast growth in the first few hours, which levels off with time. In reducing atmospheric heating, this levelling off or almost complete stop of the growing process is quite conspicuous. In either reducing or oxidizing atmosphere, a faster change is taking place at the higher temperature. Moreover, it is apparent from the figures that heating in oxidizing atmosphere brings about a very much faster growth process than does heat treatment in reducing atmosphere.

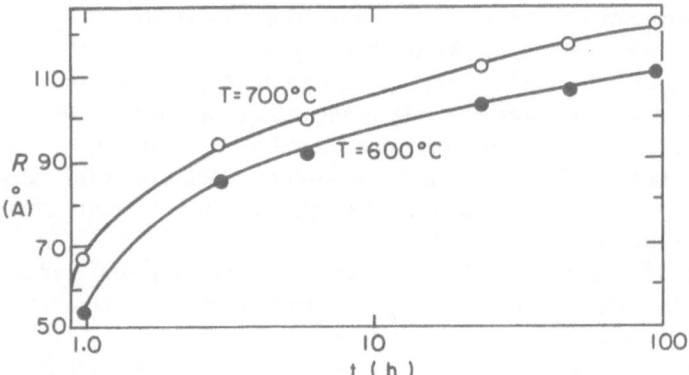

Figure 5. Change of the platinum particle size as a function of heating time in reducing atmosphere.

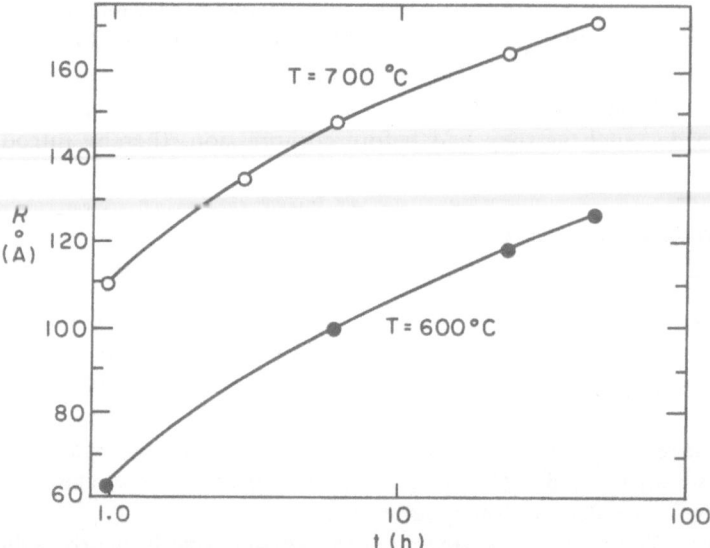

Figure 6. Change in the platinum particle size as a function of heating time in oxidizing atmosphere.

The absolute intensity of scattering (which is proportional to the number of particles whose scattering is detectable) is at first low, increases with heating time, and then eventually falls somewhat with long heating times[19]. The most straightforward interpretation would be that at first many particles are too small to be within the range of detection by the X-ray method, as heating progresses they grow larger and become detectable, and at very long times a number of them have become too large to be detectable, their scattering being hidden under the primary beam. In other words, there is no limiting particle size, but there is a steady transport of material from small particles to larger ones.

The apparent activation energy can be calculated by comparing the elapsed times for a particle to attain a given radius, at two or more temperatures, viz.,

$$E_{\text{activation}} = R \, d \ln t_r / d(1/T) \tag{4}$$

where t_r is the time to attain a given radius r.

The apparent activation energy, computed on the foregoing basis, is roughly 20 kcal for a reducing atmosphere and roughly 52 kcal for an oxidizing atmosphere. It is evident that two quite different mechanisms are involved, and it is tempting to ascribe the low-energy process to some sort of diffusion mechanism and the high-energy process to the transport through the gas phase by platinum oxide. From the work of Brewer and Elliott[25] the heat of the reaction $Pt + O_2 = PtO_2(g)$ is $\Delta H = \pm 55$ kcal, so the transport by gaseous PtO_2 becomes plausible. However, it is necessary to point out that the apparent activation energy may be relatively unreliable as a measure of the true activation energy because, as we shall show, the radius dependence is itself highly temperature dependent.

The growth or precipitation of spherical particles in a solid matrix, far from equilibrium, proceeds at a constant, mostly volumetric rate[26,27]: $r^2 dr/dt = k$, where k is the rate constant. This equation applies to relatively large particles (1 μ or more). Small particles in the 10 to 10^3 Å range due to their increased surface energy have solubilities larger than that of the large particles, as expressed by the Thompson equation[28]:

$$\ln C_r / C_s = 2V\gamma / rRT \tag{5}$$

where C_r is the solubility of particles of radius r, C_s is the limiting solubility for large particles; V is the molar volume; γ is the surface tension of the particle in the solid matrix; R is the gas constant; and T is the absolute temperature. Therefore, the rate of growth of larger particles as long as small particles are present is accelerated by a factor of $\exp(2V\gamma/rRT)$. Under these conditions the growth law becomes

$$r^2 \, dr/dt = k \, e^{a/r} \tag{6}$$

where $a = 2V\gamma/RT$. It is the presence of the surface energy term which accounts for both the decrease of rate with particle size and the increase of activation energy with particle size.

The variables of eq. (6) are easily separable, and the resulting differential equation has been integrated[19] to give

$$kt/a^3 = (r/a)^4 e^{-a/r} S(a/r) \qquad (7)$$

where the function $S(a/r)$ can be approximated by the simple equation

$$S(a/r) \simeq \frac{1 + a/r}{a/r + 5 + 3(r/a)} \qquad (8)$$

It is worth commenting that the theoretical curve is almost linear when r is plotted against log t as long as the growth is controlled by the surface energy term.

The fitting of our experimental data for the radius of gyration R to the theoretical curves (substituting R for r) requires knowledge of the parameters a and k. This can be accomplished by trial in a variety of ways.

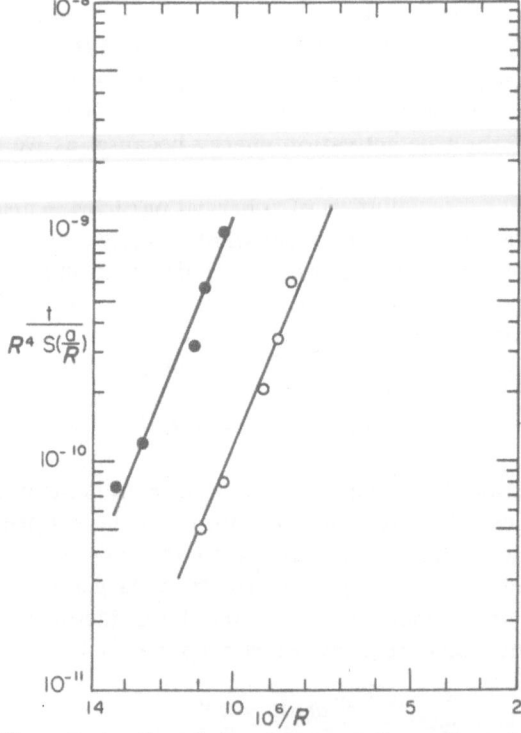

Figure 7. A plot of the change of the radius of gyration R as a function of time for heat treatment at 600°C reducing atmosphere (●) and 700°C reducing atmosphere (○).

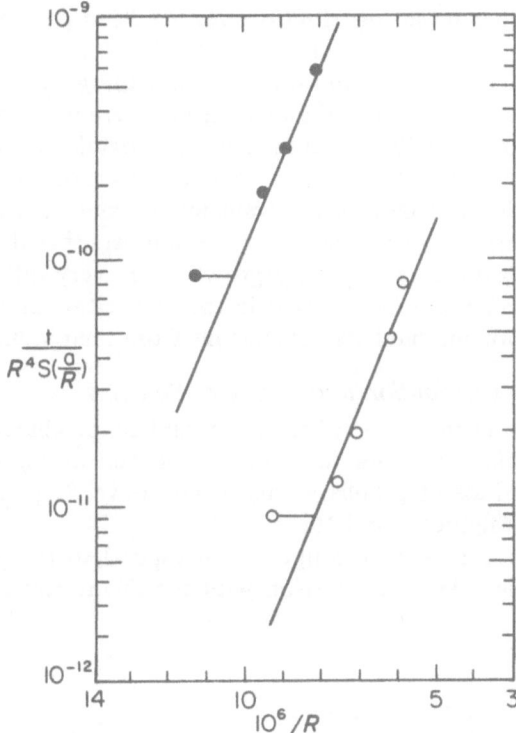

Figure 8. A plot of the change of the radius of gyration R as a function of time for heat treatment at 600°C oxidizing atmosphere (●) and 700°C oxidizing atmosphere (○).

Perhaps the simplest way is to plot $\log(t/R^4 S)$ against $1/R$, from which the slope gives a and the intercept gives ka. This kind of plot must be iterated, since S is a function of a/R and requires a prior knowledge of a; however, S is a slowly varying function and an approximate preliminary value of a permits it to be evaluated rather well[19].

The lines drawn through the data points of figures 7 and 8 were calculated from the theoretical equation with the foregoing parameters[19]. Except for the very shortest times, the agreement is within our experimental error.

The explanation for the great deviation at very short times is that it probably does not represent a real growth of (arithmetic mean) particle size at all, but represents a change in the distribution function of particles toward a broader distribution. As we shall give evidence to show later, the initial metal particles are fairly near to monodisperse, but they rapidly change to a rather broad distribution. As has been pointed out, since the X-ray method looks at particles considerably larger than the average particles, the change in distribution simulates the growth of particle size.

The rate constant k for a reducing atmosphere is given approximately by $5 \times 10^8 \exp(-38,000/RT)$ and for an oxidizing atmosphere by $2.51 \times 10^{17} \exp(-69,000/RT) \text{ Å}^3/\text{hr}$. These activation energies would be the true activation energies for the transport in contact with bulk material or very large particles. It is clear from the numbers that the two processes, reducing and oxidizing atmosphere growths, are very different. However, there is a rather large possible error in the numerical values, and it would be unsafe to draw mechanistic conclusions from their values.

5.2.2. The Size Distribution and Shape of Particles

In order to examine the distribution and shape characteristics of the platinum particles, the data have been replotted as $\log I\theta^2$ vs. $\log \theta^2$ to give the same kinds of graphs as discussed above. Representative curves are displayed in figures 9 and 10.

The dotted curves on the figures correspond to the platinum in the unheated samples. As a comparison with the theoretical curve will easily

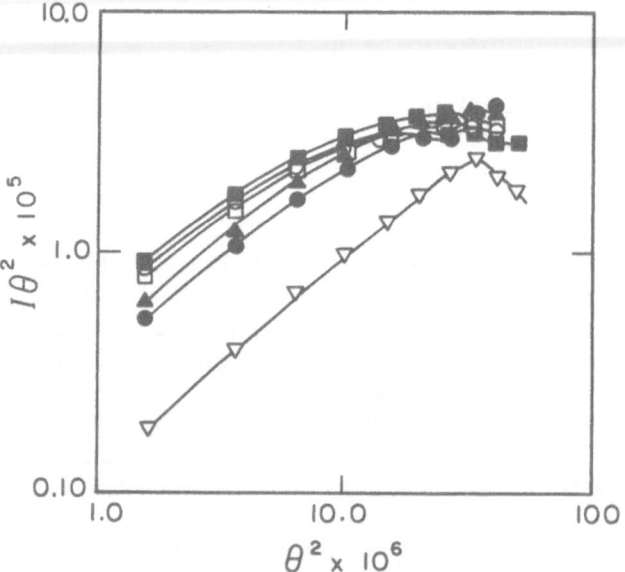

Figure 9. 5% platinum on alumina, 600°C in reducing atmosphere. ▽ heated for 1 hr or unheated; ● heated for 3 hr; ▲ heated for 6 hr; □ heated for 24 hr; ○ heated for 48 hr; ■ heated for 96 hr.

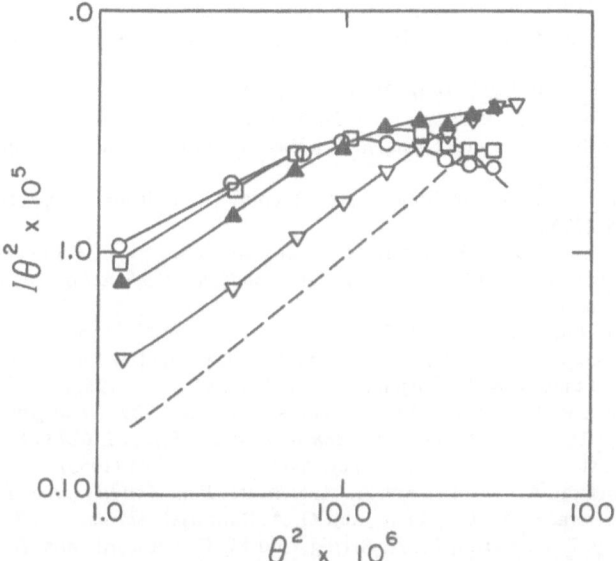

Figure 10. 5% Platinum on alumina; 600°C in oxidizing atmosphere. △ heated for 1 hr; ▲ heated for 6 hr; ☐ heated for 24 hr; ○ heated for 48 hr; — — — unheated sample.

show, these correspond rather well to the curve for monodisperse spheres; thus, it is a good approximation that the freshly deposited platinum is made up of spherical particles with a narrow distribution range.

Upon heat treatment of the samples, the scattering curves become progressively flatter, indicating a broad distribution of sizes. As time goes on, the intensity at large angles diminishes and at small angles grows, which of course indicates that smaller particles are disappearing while large particles are growing. The curves are similar in shape to those for a Maxwellian distribution of spheres with $n = 3$ or less in $N(r)$.

The very flat distribution is particularly noticeable in the data for an oxidizing atmosphere. The theoretical curves for a flat distribution of spheres are rather similar to those for a distribution of discs, and it is not possible to make a firm decision as to the shape. It is not implausible that in the oxidizing atmosphere the particles redeposit on the alumina in a somewhat platelike form since platinum oxide may well be able to wet alumina.

REFERENCES

1. A. Guinier and G. Fournet, *Small-Angle Scattering of X-Rays*, John Wiley, New York (1955).
2. J. A. Simpson, *Rev. Sci. Instr.* **35**, 1698 (1964).
3. C. W. Caldwell, Jr., *Rev. Sci. Instr.* **36**, 1500 (1965).
4. B. K. Vainshtein, Structure Analysis by Electron Diffraction, Pergamon Press, New York (1964).
5. J. J. Lander, Progress in Solid State Chemistry, (H. Reiss, ed.), Pergamon Press, New York (1965).
6. A. Chutjian and G. A. Somorjai, University of California, UCRL-17126 (1967).
7. R. E. Schlier and H. E. Farnsworth, *Semiconductor Surface Physics*, University of Pennsylvania Press, Philadelphia (1956).
8. A. U. MacRae, *Surface Sci.* **4**, 247 (1966).
9. N. A. Gjostein, *Metal Surfaces*, Chap. IV, American Society for Metals (1963).
10. H. B. Lyon and G. A. Somorjai, *J. Chem. Phys.* **44**, 3707 (1966).
11. B. C. Clark, R. Herman, and R. F. Wallis, *Phys. Rev.* **139**, A860 (1965).
12. A. U. MacRae and L. H. Germer, *Ann.N.Y. Acad. Sci.* **101**, 627 (1963).
13. L. G. Feinstein and D. P. Shoemaker, *Surface Sci.* **3**, 294 (1965).
14. G. A. Somorjai, A. Chutjian, and A. M. Mattera, *Bull. Am. Phys. Soc.* **11**, 727 (1966).
15. A. M. Mattera, R. M. Goodman, and G. A. Somorjai, *Surface Sci.* **7**, 26 (1967).
16. H. A. Levy, P. A. Agron, M. A. Bredig, and M. D. Danford, *Ann. N.Y. Acad. Sci.* **79**, 762 (1960).
17. N. W. Gingrich, *Liquids; Structure, Properties, Solid Interactions*, (T. J. Hughel, ed.), American Elsevier, New York (1965).
18. H. B. Lyon and G. A. Somorjai, *J. Chem. Phys.* **46**, 2539 (1967).
19. G. A. Somorjai, PhD thesis, University of California (1960).
20. W. Parrish and T. R. Kohler, *Rev. Sci. Instr.* **27**, 795 (1956).
21. L. C. Roess and C. G. Shull, *J. Appl. Phys.* **18**, 295 (1947).
22. E. A. Porai-Koshits and L. S. Andreyev, *J. Soc. Glass Technol.* **43**, 235T (1959).
23. E. L. Gunn, Meeting of the American Chemical Society, Division of Petroleum Chemistry, April 1958.
24. P. W. Montgomery, H. Stromberg, and G. Jura, University of California UCRL-9796 (1961).
25. L. Brewer and G. R. Elliott, PhD thesis, University of California (1952).
26. F. J. Morin and H. Reiss, *J. Phys. Chem. Solids* **3**, 196 (1957).
27. G. A. Somorjai, *J. Chem. Phys.* **35**, 655 (1961).
28. H. K. Hardy and T. J. Heal, *Progr. Metal Phys.* **5**, 143 (1954).

VII. THE ANALYSIS OF LOW-ANGLE LIGHT SCATTERING FROM SIMPLE MIXTURES*

Daniel Caulfield

Forest Products Laboratory
Madison, Wisconsin

and

Yung-Fang Yao and Robert Ullman

Ford Motor Company, Scientific Laboratory
Dearborn, Michigan

The scattering of light is a well-known tool for the determination of surface structure, particle shape, and particle size. The equations governing the physical phenomenon are similar to those used in X-ray scattering; the wavelength of light in a scattering medium is about 2000 times the wavelength of X-rays commonly used. Accordingly, the scattering of light tends to be useful for studies of physical dimensions larger than those easily treated in X-ray studies.

An important problem is that the refractive index at visible wavelengths varies considerably with material composition, which is not common in X-ray investigations. Accordingly, the simple theory used normally in X-rays and light scattering (the oscillating dipole), breaks down in certain refractive index and particle size ranges. This constitutes an important limitation on the practical applicability of light-scattering studies.

We consider here light-scattering patterns from suspension of spheres imbedded in solid matrices. The materials used are spheres of glass and polystyrene lattices in various plastic media. Measurements were made between 10 min. and 20°. The interpretation of the experiments in terms of simple models is discussed.

One of the fascinating aspects of this problem has to do with the behavior of the derivative of the correlation function, $\gamma(r)$, in the neighborhood of $r = 0$. Specific surface is proportional to this quantity, but a mathematical discontinuity associated with the nominal scattering at experimentally inaccessible angles introduces difficulties which may lead to errors in interpretation of the experiment. The consequences of this problem are examined.

---

* In partial fulfillment of the requirements of the PhD degree of D. Caulfield at the Polytechnic Institute of Brooklyn, New York.

1. INTRODUCTION

The results of the research presented here are largely incomplete and are unsuccessful in the sense that the work was directed toward specific objectives. Our attempts to carry out meaningful experiments were nevertheless highly instructive to us, and it seemed worthwhile to set down some of the problems and our efforts to treat them together with other presentations in a symposium in which small-angle scattering plays a prominent role.

The research originated following a report written by Peter Debye to the American Petroleum Institute entitled "Characterization of the Physical Structure of Porous Materials by an Optical Method"[1]. The basic ideas in Debye's report have been published elsewhere[2][3] and deal with light or X-ray scattering as a means of structure determination of noncrystalline materials. It is known that it is possible, in principle, to determine particle sizes, surface areas, and other characteristic dimensions of a system by an analysis of the scattering pattern. Indeed, any kind of structural parameter derivable from a pair distribution function may be deduced, in principle, from a properly designed set of experiments. A detailed treatment of this is presented in a treatise by Guinier and Fournet[4]. It was pointed out by Debye and independently by Porod[5] that if a two-component mixture is random in a certain sense the correlation function takes on a simple form, an exponential characterized by a single parameter. For systems which could be classified in this way, a very complete structural description was possible.

The interest of the American Petroleum Institute in this area of research centered on the very practical problem of getting oil out of the ground. Oil is formed in porous rock structures; the quantity of oil which can be efficiently removed from these rocks is intimately related to the microscopic geometry of the channels in the rock. Any procedure which could shed further light on the structure of porous minerals might be worthy of careful investigation. If the structural parameters obtained could be correlated with the ability to move oil out of a rock mass, the method could be quite useful. Debye suggested that light-scattering studies might conceivably be used to obtain a better characterization of the geometry of these materials. The particle size of oil-bearing minerals can vary widely from the submicron range to objects macroscopic in size (1 mm or larger). It is evident that the wavelength of visible light is a suitable yardstick for such a study provided that the scattering measurement is carried out at low angles. The angular range most essential for effective characterization of a sample depends on the material under study: low angles for large particles and vice versa.

It became apparent to us after some trial experiments with sections of rock samples that it would be wise to concentrate on the study of model systems at the beginning rather than tackle the problem of working with real rock sections. Accordingly, most of the scattering experiments were carried out on suspensions of glass beads and spheres of polystyrene latex in rigid media. These were chosen because light scattering and microscopic measurements could be easily compared if the suspensions were of simple geometry. In addition, the surface areas of glass beads obtained by gas adsorption could be used for further comparison with light-scattering results.

2. BASIC SCATTERING THEORY

Light is scattered because the electromagnetic waves of the light beam generate a high-frequency field which induces electric oscillation in the material. If the material is transparent, the electrons oscillate in phase with the incident wave and generate a radiation field of the same frequency as the incident light. The moving electrons are coupled to positively charged nuclei and act as oscillating dipoles, quadrupoles, octopoles, etc. The scattered light may be observed at various angles to the incident beam.

The exact relationships between the angular pattern of scattered light and the form of the scattering particles are complex and, with the exceptions of spheres, infinite cylinders, and a few other figures of simple geometry, have not been thoroughly worked out. Fortunately, for a great number of experimental situations, where the variation of refractive index in the scattering sample is not too large and where spatial correlations are not too long range, the scattering pattern may be calculated by using only the electric dipole terms of the radiation field, which is the leading term in the multipole expansion. The scattering which originates from oscillating electric dipoles is commonly referred to as Rayleigh–Gans (RG) scattering.* The analysis of a scattering pattern for a sample whose material properties are consistent with the RG criteria leads to useful information about the particle structure and is not limited to particle shapes of simple geometry. While the experiments of this investigation are carried out with suspensions of spheres for which a scattering theory is available for all possible combinations of particle size and refractive index[6], the ultimate purpose is to treat porous minerals of complicated geometry where only the RG analysis is available. Accordingly, the interpretation of the data is limited to the RG treatment (this is formally identical with the theory of X-ray scattering), and the experiments are designed with this limitation in mind.

Consider a parallel light beam passing through a scattering sample of frequency ω with an electric vector $\mathbf{E}^0$ (the direction of polarization). The

---

* A general classification of light scattering as a function of particle size, refractive index, and absorption properties will not be given here; readers are referred to the classic treatise of van der Hulst[6] for a complete categorization of this problem.

observer of the scattered light is at a position r_0; the electric vector at r_0 is composed of the scattered waves originating from all points in the scattering sample. The angle between the incident light wave and the scattered light is θ. The frequency of the scattered light is the same as the incident light; electric vector at r_0 is obtained by summing over all points in the scattering volume. The equations which link the electric field and intensities of the incident and scattered light are

$$\mathbf{E}^0(\mathbf{r}_i) = \mathbf{E}_0{}^0 \exp(i\omega t - \mathbf{k}_0 \cdot \mathbf{r}_i) \tag{1a}$$

$$\mathbf{E}^0(\mathbf{r}_i; \mathbf{r}_0) = \mathbf{A} \cdot E_0{}^0 \exp[i(\omega t - \mathbf{k}_0 \cdot \mathbf{r}_i - \mathbf{k} \cdot r_{i0}] \tag{1b}$$

$$\mathbf{E}(\mathbf{r}_0) = \sum_i \mathbf{E}(\mathbf{r}_i; \mathbf{r}_0) \qquad \text{(discrete array of scattering centers)} \tag{1c}$$

$$\mathbf{E}(r_0) = \int E(\mathbf{r}_i; \mathbf{r}_0) dr_i \qquad \text{(continuous distribution of scattering centers)} \tag{1d}$$

$$I_0 = \frac{c}{8\pi} \langle \mathbf{E}^0(\mathbf{r}_i) \cdot E^{0*}(\mathbf{r}_i) \rangle_t \tag{1e}$$

$$I(\mathbf{r}_0) = \frac{c}{8\pi} \langle \mathbf{E}(\mathbf{r}_0) \cdot \mathbf{E}^*(r_0) \rangle_t \tag{1f}$$

where $\mathbf{E}_0{}^0$ is the amplitude of the electric vector of the incident light wave; $E^0(\mathbf{r}_i)$ is the electric wave at a point r_i; $\mathbf{E}(\mathbf{r}_i; \mathbf{r}_0)$ is the electric vector at the position of an observer created by a scattering unit at r_i; $E(\mathbf{r}_0)$ is the total electric vector of the scattered wave at r_0; I_0 is the intensity of the incident light beam per square centimeter; $I(\mathbf{r}_0)$ is the intensity of the scattered beam at r_0 per unit solid angle; $\langle \rangle_t$ denotes a time average; c denotes speed of light; $\mathbf{k}_0$ and $\mathbf{k}$ are wave vectors in the direction of the incident and scattered waves respectively and are equal in magnitude to $2\pi/\lambda$; $\mathbf{A}$ is a second-order tensor which is equal to a constant multiplied by the polarizability tensor minus those components which generate an electric field vector in the direction of the scattered light—specifically

$$\mathbf{A} = \frac{4\pi^2}{\lambda_0{}^2 r_{i0}} (\boldsymbol{\alpha} - (\mathbf{s} \cdot \boldsymbol{\alpha})\mathbf{s}) \tag{2}$$

λ_0 is the wavelength of the light in vacuum; $\mathbf{s}$ is a unit vector in the direction of the scattered light which equals $\mathbf{k}/\|\mathbf{k}\|$; $\boldsymbol{\alpha}$ is the polarizability tensor per particle [eq. (1b)] or per unit volume [eq. (1c)], depending on whether a discrete or continuous array of scattering units is envisaged.

In many simple cases the induced oscillating dipole is parallel to the incident field of the light wave, in which case **A** may be written as

$$\mathbf{A} = \frac{4\pi^2}{\lambda_0{}^2 r_{10}} \alpha \mathbf{1} \qquad \text{(vertically polarized light)} \qquad (3a)$$

$$\mathbf{A} = \frac{4\pi^2 \alpha \cos\theta}{\lambda_0{}^2 r_{10}} \mathbf{1} \qquad \text{(horizontally polarized light)} \qquad (3b)$$

In the subsequent analysis it will be assumed that the light is vertically polarized and that eq. (3a) is applicable. From this one may write

$$I(\mathbf{r}_0) = \frac{16\pi^4}{\lambda_0{}^4 r_0{}^2} I_0 \sum_{i,j} \alpha_i \alpha_j \exp(i\mathbf{h}\cdot\mathbf{r}_{ij}) \qquad (4a)$$

or

$$I(\mathbf{r}_0) = \frac{16\pi^4}{\lambda_0{}^4 r_0{}^2} I_0 \int\int \alpha(\mathbf{r}_i)\alpha(\mathbf{r}_j) \exp[i\mathbf{h}\cdot\mathbf{r}_{ij}]d\mathbf{r}_i d\mathbf{r}, \qquad (4b)$$

$$\mathbf{h} = \mathbf{k} - \mathbf{k}_0$$

$$\mathbf{h} = |\mathbf{h}| = \frac{4\pi}{\lambda}\sin(\theta/2)$$

$(4c)$

where λ, the wavelength of the light in medium of refractive index n, equals λ_0/n.

The oscillatory property of the term $\exp(i\mathbf{h}\cdot\mathbf{r}_{ij})$ in eqs. (4a) and (4b) would cause $I(\mathbf{r}_0)$ to vanish in any scattering sample of macroscopic size if $\alpha(\mathbf{r}_i)$ were independent of position and, in fact, if one writes

$$\Delta\alpha(\mathbf{r}_i) = \alpha(\mathbf{r}_i) - \bar{\alpha} \qquad (5a)$$

$$\bar{\alpha} = \frac{1}{V}\int \alpha(\mathbf{r}_i)\, d\mathbf{r}_i \qquad (5b)$$

the scattering of eqs. (4a) and (4b) can be entirely accounted for by the fluctuations in polarizability. The polarizability per unit volume is related to refractive index by

$$\alpha(\mathbf{r}_i) = \frac{n^2 - 1}{4\pi} \qquad (6a)$$

$$\Delta\alpha(\mathbf{r}_i) = \frac{n\Delta n(\mathbf{r}_i)}{2\pi} \qquad (6b)$$

while the polarizability per particle is linked to refractive index through the relationship $\alpha_i = V/N \, \alpha(r_i)$, where N is the number of scattering particles in the volume V. By substitution, eqs. (4a) and (4b) become

$$i(\mathbf{h}) = \frac{4\pi^2 n^2 V}{\lambda_0^4 N^2} \sum_{i,j} \Delta n_i \Delta n_j \exp(i\mathbf{h} \cdot \mathbf{r}_{ij}) \tag{7a}$$

$$i(\mathbf{h}) = \frac{4\pi^2 n^2}{\lambda_0^4 V} \int \int \Delta n(\mathbf{r}_i) \Delta n(\mathbf{r}_j) \exp(i\mathbf{h} \cdot \mathbf{r}_{ij}) d\mathbf{r}_i \, d\mathbf{r}_j \tag{7b}$$

$$i(\mathbf{h}) = \frac{r_0^2 I(\mathbf{r}_0)}{I_0 V} \tag{7c}$$

where $i(\mathbf{h})$ is the Rayleigh ratio or reduced scattering function, the relative intensity of scattered light per volume per unit solid angle.

Equation (7b) may be put in a convenient form by definition of an autocorrelation function

$$\gamma(\mathbf{r}) = \frac{\int \Delta n(\mathbf{R}) \Delta n(\mathbf{R} + \mathbf{r}) \, d\mathbf{R}}{\int [\Delta n(\mathbf{R})]^2 \, d\mathbf{R}} \tag{8a}$$

$$\overline{(\Delta n)^2} = \frac{1}{V} \int [\Delta n(\mathbf{R})]^2 \, dR \tag{8b}$$

Setting $\mathbf{r}_i = \mathbf{R}$ and $\mathbf{r}_j = \mathbf{R} + \mathbf{r}$ eq. (7b) becomes

$$i(h) = \frac{4\pi^2 n^2}{\lambda_0^4} \overline{(\Delta n)^2} \int_0^\infty \gamma(\mathbf{r}) e^{i\mathbf{h} \cdot \mathbf{r}} \, d\mathbf{r} \tag{9a}$$

For isotropic systems where the autocorrelation function depends on the separation between two points only, $\gamma(\mathbf{r}) = 4\pi r^2 \gamma(r) \, dr$ and eq. (9a) reduces to

$$i(h) = \frac{16\pi^3}{\lambda_0^4} n^2 \overline{(\Delta n)^2} \int_0^\infty r^2 \gamma(r) \frac{\sin hr}{hr} \, dr \tag{9b}$$

The correlation function may be obtained in principle from eqs. (9a) and (9b) by a Fourier transform;

$$\gamma(\mathbf{r}) = \frac{\lambda_0^4}{32\pi^5 n^2 \overline{(\Delta n)^2}} \int i(\mathbf{h}) e^{i\mathbf{h} \cdot \mathbf{r}} \, d\mathbf{r} \tag{10a}$$

which for an isotropic sample becomes

$$\gamma(r) = \frac{\lambda_0^4}{8\pi^4 n^2 \overline{(\Delta n)^2}} \int_0^\infty h^2 i(h) \frac{\sin hr}{hr} \, dh \tag{10b}$$

While eqs. (7b), (8), (9), and (10) are useful in dealing with continuous distributions of matter, eq. (7a) is often convenient in dealing with the excess scattering from a set of particles in a continuous medium. By averaging over all orientations for an isotropic system, one finds eq. (7a) to be

$$i(h) = \frac{16\pi^3 n^2 (\Delta n)^2 V}{\lambda_0^4 N^2} \sum_{i,j=1}^{N} \frac{\sin hr_{ij}}{hr_{ij}} \qquad (11)$$

The further division of the N scattering particles into aggregates, ν subunits per aggregate, and m aggregates ($m\nu = N$), leads to simplification of eq. (11), provided that the m aggregates are sufficiently far apart to be independent scatterers. Then eq. (11) reduces to

$$i(h) = \frac{16\pi^3 n^2 (\Delta n)^2 M}{\lambda_0^4 N_A c} \frac{1}{\nu^2} \sum_{i,j=1}^{\nu} \frac{\sin hr_{ij}}{hr_{ij}} \qquad (12a)$$

where M is the molecular weight of an aggregate; N_A is Avogadro's number; and c is the concentration in g/ml. The increment in refractive index is nearly always directly proportional to concentration, in which case $(\Delta n)^2 = (\delta n/\delta c)^2 c^2$ and eq. (12a) becomes

$$i(h) = \frac{16\pi^3 n^2 (\partial n/\partial c)^2 Mc}{\lambda_0^4 N_A} \frac{1}{\nu^2} \sum_{i,j=1}^{\nu} \frac{\sin hr_{ij}}{hr_{ij}} \qquad (12b)$$

The extension of eqs. (12a) and (12b) to include scattering from assemblies of aggregates whose positions are correlated may be represented by including another term summing over $\sin hr_{ij}/hr_{ij}$ where the scattering centers i and j are not in the same aggregate. This can only be carried through in detail if the spatial distribution of aggregates is known. In referring this to the light scattering of solutions, one obtains virial expansions in which the virial coefficients are functions of scattering angle.

In the representation of scattering from a continuous distribution of matter, the formulas for absolute scattering intensities may be presented in a different (and sometimes more convenient) form by using the fact that $\gamma(0) = 1$ (see eq. 8a). From this and eq. (10b) one can write

$$\gamma(r) = \int_0^\infty h^2 i(h) \frac{\sin hr}{hr} \, dh \Big/ \int_0^\infty h^2 i(h) \, dh \qquad (13a)$$

from which it follows that, by comparison with eq. (9a),

$$i(h) = \frac{2}{\pi} \left(\int_0^\infty h^2 [i(h) \, dh] \right) \left(\int_0^\infty r^2 \gamma(r) \frac{\sin hr}{hr} \, dr \right) \qquad (13b)$$

3. DETERMINATION OF CHARACTERISTIC DIMENSIONS

3.1. Radius of Gyration

For a system of widely separated particles, the scattering may be related to dimensions of the particles by eq. (12a). Expanding in a Taylor series, one obtains

$$i(h) = \frac{A}{\nu^2} \sum_{i,j} \left(1 - \frac{h^2 r_{ij}^2}{6} + O(h^4) \right) \tag{14}$$

where A is a lumped constant. If the experiments are carried out at sufficiently low scattering angles so that terms of the order of h^4 may be ignored (the larger the particle size, the lower the scattering angle), one finds

$$i(h) \approx A\, i(0) \left(1 - \frac{h^2 R^2}{3} \right) \approx i(0) \exp(-h^2 R^2/3) \tag{15a}$$

$$R^2 = \frac{1}{\nu} \sum_{j=1}^{\nu} \mathbf{r}_{0j}^2 = \left(\frac{1}{2\nu^2} \right) \sum_{i,j=1}^{\nu} r_{ij}^2 \tag{15b}$$

where $\mathbf{r}_{0j}$ is the vector connecting the centre of mass with the jth segment; and R is the radius of gyration; and $i(0)$ is the theoretical Rayleigh ratio at zero angle. Equation (15a) is known as Guinier's equation. The radius of gyration may be obtained from a plot of $i(h)$ vs. h^2 or $\log i(h)$ vs. h^2.

3.2. Correlation Function

It should be kept in mind that a meaningful measurement of radius of gyration of a particle can only be obtained for systems of widely separated particles. For certain samples in which the particulate nature of the system is not clearly defined, alternative approaches using the correlation function are applicable. The physical sense of the use of the correlation function is more easily perceived if the system is a mixture made up of two components, each of which is internally homogeneous. This means that the fluctuations in refractive index are entirely determined by the distribution of the two components in the sample. If the refractive indices of the two materials are n_1 and n_2, the corresponding volume fractions are φ_1 and φ_2, the refractive index is given by

$$n = n_1 \varphi_1 + n_2 \varphi_2 \tag{16a}$$

and the mean-square fluctuation in refractive index is given by

$$\overline{(\Delta n)^2} = \varphi_1 \varphi_2 (n_1 - n_2)^2 \tag{16b}$$

The correlation function defined in eq. (8a) may be written

$$\gamma(r) = \frac{\overline{(\Delta n_A \Delta n_B)}}{\overline{(\Delta n)^2}} \tag{17}$$

where Δn_A and Δn_B are the fluctuations in refractive index at points A and B which are separated by a distance r. $\gamma(0)$ is unity and $\lim_{r \to \infty} \gamma(r) = 0$ since the fluctuations in refractive index are uncorrelated at large distances.

3.3. Distance of Heterogeneity

A useful structural parameter is l_c, the distance of heterogeneity being defined by

$$l_c = 2 \int_0^\infty \gamma(r)\, dr = \left(\pi \int_0^\infty h i(h)\, dh \right) \Big/ \left(\int_0^\infty h^2 i(h)\, dh \right) \tag{18}$$

In a system of widely separated particles, l_c is twice the mean length of all chords joining two points in the particle and contained within the particle taken in all possible directions([4]). For systems of more complicated geometry, l_c is a weighted average of segments of a straight line passing through the sample averaged over all possible straight lines. A line segment in such a system is a length of line which lies entirely within a single component of the mixture.

3.4. Surface Area

A most interesting application of the correlation function is the determination of surface area. It is based on the notion that if a stick of length r is placed in a two-component mixture with one end embedded in the first component, the probability that the other end of the stick is also embedded in the first component is equal to $\gamma(r)$. If r becomes very small, the difference between $\gamma(r)$ and $\gamma(0)$ is proportional to the probability that the first end is very close to an interface; therefore, this difference is proportional to the surface area. In fact, a series expansion of $\gamma(r)$ leads to the result

$$\gamma(r) = 1 - \frac{S}{4\varphi_1\varphi_2 V} r + O(r^2) \tag{19a}$$

$$S/V = -4\varphi_1\varphi_2\gamma'(0) \tag{19b}$$

The initial slope of the correlation function yields the specific surface. Some difficulties with this formula arising because of unavailability of data at large values of h will be treated later. It is possible to eliminate the determination of the correlation function and obtain the specific surface of a material by direct examination of the high-angle limit of the low-angle scattering curve. By integration of eq. (13b) after substitution of eq. (19a) for $\gamma(r)$ one obtains

$$I(h) \simeq \frac{S}{h^4 \varphi_1 \varphi_2 V} \int_0^\infty h^2 I(h)\, dh \tag{20}$$

A plot of $h^4 I(h)$ vs. h may be used to calculate surface area. This function contains damped oscillatory terms, and the surface area can be obtained if there is a range of h where $h^4 I(h)$ remains essentially constant. This result is modified if the light source does not emanate from a single point. For a light source which is a thin "infinitely long" line, it turns out that $h^3 I(h)$ is proportional to the specific surface. For sources which may be approximated by neither a point nor an infinite line, analysis of the data is more intricate.

It should be stressed that the correlation function $\gamma(r)$, the radius of gyration R, the distance of heterogeneity l_c, and the specific surface S/V in principle can be obtained unambiguously from the intensity of scattered light as a function of angle without the introduction of any additional hypotheses (except those already mentioned) about the nature, distribution, or arrangement of the scattering elements. If certain additional assumptions are introduced, simplifications ensue. The validity of the results so obtained depends upon the correctness of the extra assumptions.

3.5. The Debye Approximation

Debye[3] has shown that a sample which has a random distribution of surface elements yields a correlation function of a simple form,

$$\gamma(r) = e^{-r/a} \tag{21}$$

In this expression, a has the dimensions of length and has been called the correlation distance. The form of this correlation function meets the general requirement that $\gamma(r) = 1$ at $r = 0$ and γ approaches zero as r becomes very large. The correlation distance is just one half of the distance of heterogeneity, and the relation between surface and correlation distance is given by [see eq. (19b)]

$$\frac{S}{V} = \frac{4\phi_1 \phi_2}{a} \tag{22}$$

When the Debye correlation function is introduced into the expression for the intensity as a function of h, an integration may be performed directly to yield

$$i(h) \sim \frac{a^3}{(1 + h^2 a^2)^2} \tag{23}$$

It is clear that, by means of this expression, an experimental determination of the function $I(h)$ affords a way of evaluating the correlation distance (by plotting $[i(h)]^{-\frac{1}{2}}$ vs. h^2) and thereby the specific surface of the sample.

3.6. Particle Size Distribution

The evaluation of surface area in principle also can be obtained through radius of gyration measurements in systems of widely dispersed particles. However, the significance of the experimentally determined radius of gyration is complicated by the fact that rarely does one encounter a system of uniformly sized particles. Instead, the scattering particles may conform to any of an infinite number of distributions of particle sizes. In a procedure suggested by Hosemann[7], the weight fraction of particles having a radius of gyration between R and $R + dR$ is assumed to be represented by a generalized exponential function. If the particles conform to this type of distribution, a mean radius of gyration may be determined. However, since the radius of gyration, obtained from a scattering experiment, is indepen- dent of particle shape, its determination does not provide by itself a measure of the surface of the scattering sample. If one knows the form of the particles and can determine the distribution of particle sizes, then in principle one could calculate the surface of the scattering sample. Many suggestions have been made for obtaining particle-size distributions. Shull and Roess[8] proposed a method similar to Hosemann's in which they too assume a general form for the distributions. Jellinek, Solomon, and Fan- kuchen[9] have demonstrated a geometric technique for obtaining an approximate particle-size distribution. Bauer[10], Roess[11], and Riseman[12] have proposed methods of deriving the exact particle-size distribution for spheres from scattering data by analytic means. In 1957 Luzzati[13] pro- posed a general scheme for obtaining the size distribution of particles analytically by using the properties of the Fourier integral. This is based on the following argument.

Consider a sample composed of N homothetic (each particle has the same shape) particles where $n(R)$ is the fraction of particles having a certain characteristic dimension equal R. If $i(hR)$ is the scattering function per particle, Roess[11] has shown that

$$I(h) = N \int_0^\infty R^6 i(hR) n(R) \, dR \qquad (24)$$

Equation (24) has the form

$$f(x) = \int_0^\infty k(xy) g(y) \, dy \qquad (25)$$

where $f(x)$ and $k(xy)$ are known functions. Let $x = e^t$ and $y = e^{-v}$.

Equation (25) takes the form

$$\varphi(t) = \int_{-\infty}^{\infty} q(t - v)\gamma(v)\,e^{-v}\,dv = \int_{-\infty}^{\infty} q(t - v)\beta(v)\,dv \qquad (26)$$

Using the convolution theorem of the Fourier transform one has

$$F(s) = K(s)B(s) \qquad (27)$$

where $F(s)$, $K(s)$ and $B(s)$ are the Fourier transforms of $\varphi(t)$, $q(t)$, and $\beta(t)$, respectively. $B(s)$, unknown, is obtained from eq. 27. By using the inverse transform one obtains $\beta(v)$, $\gamma(v)$, and $g(y)$, the latter of which represents $R^6 n(R)$ in eq. (24).

This very ingenious technique has limited utility, however. First, since it is rare that all scattering particles in a sample have the same shape; second, even if the particles did have the same shape, the precision of scattering measurements is such that relatively small errors in the determination of scattering intensity leads to large errors in the calculation of size distribution. There has not been, to our knowledge, any systematic analysis of the limitations of this technique by an analysis of how the errors of measurement of scattering intensity are quantitatively related to the accuracy of the particle-size distribution determined therefrom.

3.7. Multiple Scattering

A major difficulty encountered in light-scattering studies of strong scatterers is that of multiple scattering. In most analyses of experimental data it is assumed that the light beam is only scattered once. If this is not correct, the deductions from the scattering pattern will be wrong. In general, the best method of dealing with the problem of multiple scattering is to recognize its seriousness and to attempt to avoid it. The best indication of multiple scattering comes from a measurement of the transmission of the scattering sample. In some cases this is difficult because of the large intensity of light scattered in the forward direction. One can estimate, however, that if the transmission of the scattering sample is 90%, the multiply scattered light makes a contribution to the scattered intensity of the order of 1%. The three general remedies for multiple scattering are (1) dilution of the scattering sample, (2) decreasing the thickness of the sample, and (3) closer matching of the refractive indices of scattering particles and surrounding matrix. The choice of the remedy depends largely upon the type of sample under investigation.

Luzzati[14] and Soulé[15] have shown how multiple scattering may be put to use in scattering experiments provided that the relationship $\sin \theta \approx \theta$

is valid throughout the scattering and multiple scattering range. The multiple scattering pattern may be represented theoretically by repeated use of the convolution theorem of the Fourier integral. This is particularly adaptable to the calculation of surface areas and has been applied to the determination of surface area of finely divided carbon and graphite by X-ray scattering[16].

There is one special circumstance in which the study of multiple scattering patterns may be an advantage. Multiple scattering extends to higher angles than a single scattering process. Where the scattering is so close to the incident beam that data is difficult to collect, the broadening due to multiple scattering is useful. Otherwise, it is best to eliminate multiple scattering as described above.

3.8. Effect of the High-Angle Limit on Surface Area Calculations

It is evident from the representation of the correlation function $\gamma(r)$ as a Fourier transform of the reduced scattering function $i(h)$—see eq. (10b)—that scattering data must be tabulated as h takes on all values from zero to infinity. Experimentally, one is limited at low angle by the need to separate the scattered light from the direct beam; at high angle one is limited either by the fact that the scattering decreases to a level which cannot be separated from the background or by the experimental upper limit on $h = 4\pi\lambda \sin(\theta/2)$ which is $4\pi/\lambda$.

The high-angle limit introduces a serious theoretical problem in surface area determination[17]. This is best described as follows. According to eq. (19b), the specific surface is proportional to $\gamma'(0)$, the initial slope of the correlation function. If the infinite upper limit in eq. (10b) is replaced by h^*, which is less than or equal to $4\pi/\lambda$, one has

$$\gamma_{exp}(r) \sim \int_0^{h*} h^2 I(h) \frac{\sin hr}{hr} dh \qquad (28)$$

The subscript "exp" refers to the correlation function calculated from data with an experimental upper limit h^*.

Expanding $\sin hr/hr$ in a power series in h, convergent for all finite values of h, one has

$$\gamma_{exp}(r) = 1 - \gamma_1 r^2 + \gamma_2 r^4 + \ldots \qquad (29a)$$

from which it follows that

$$\gamma'_{exp}(0) = 0 \qquad (29b)$$

The difference between this result and eq. (19b) arises because $\gamma'(r)$ is nonuniformly convergent for small values of r. An exact measure of $\gamma'_{exp}(r)$ for a dilute suspension of spheres has been treated elsewhere[17], together with a detailed discussion of this problem. Some results are shown in figure 1. It is evident that a judicious choice of the small value of r for estimating $\gamma'(r)$ near but not at the limit of $r = 0$ can lead to approximate values of specific surface that are reasonable. The range of h required in order that eq. (20) be useful needs investigation.

3.9. The Assumption of Random Surface Elements

The treatment of Debye *et al.* leading to $\gamma(r) = e^{-r/a}$ for a random two-component mixture permits the calculation of surface areas and particle size in terms of the single parameter a. For a dilute suspension the specific surface of the suspended particles is given by

$$S_{\mathrm{sp}} = 4/a \qquad (30)$$

This is obtained from eq. (22), recognizing that $S_{\mathrm{sp}} = S/(\varphi_1 V)$ and $\varphi_2 = 1$. From eq. (23) one may write

$$i(h) \sim a^3[1 - 2h^2a^2 + 0(h^4)] \qquad (31a)$$

$$a^2 = R^2/6 \qquad (31b)$$

Equation (31b) is obtained by comparison of (15a) and (31a).

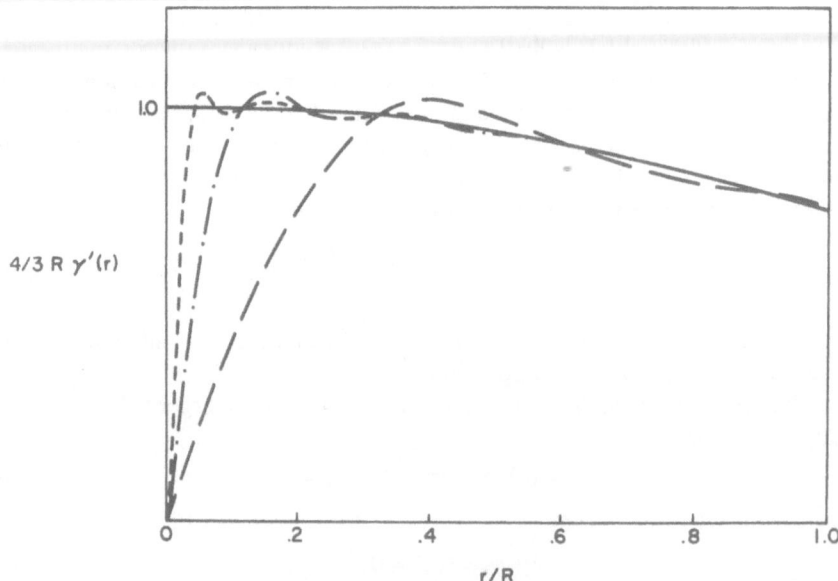

Figure 1. A plot of $-4/3 R\gamma'(r)$ vs. r/R for a dilute assembly of spheres of radius R for varying values of h^* [eq. (28)] ——————, $h^*R = \infty$; – – – –, $h^*R = 90$; —·—·—·, $h^*R = 30$; —————, $h^*R = 10$.

Equation (30) and (31b) are the logical consequence of the use of the correlation function $\gamma(r) = e^{-r/a}$ for scattering studies. It is instructive to determine a form for a suspension of spheres by using eq. (30) and (31b) independently in order to estimate the errors possibly made by the assumption $\gamma(r) = e^{-r/a}$.

From eq. (30) one obtains for spheres

$$a^2 = (16/9)R_0{}^2 \tag{32a}$$

R_0 is the radius of the sphere. This result originates from the high-angle limit of the scattering pattern.

From eq. (31b), using the fact that $R^2 = \frac{3}{5}R_0{}^2$, one finds

$$a^2 = R_0{}^2/10 \tag{32b}$$

a result which follows from the scattering at low angles.

The inconsistency between eqs. (32a) and (32b) is very great, which might be expected for dilute suspension of spheres where surface elements are by no means randomly placed. From the magnitude of the discrepancy, however, it is evident that an arbitrary selection of the form of the correlation function may have a considerable effect on the apparent values of the parameters of the system deduced from scattering measurements.

3.10. Limitations of the Rayleigh–Gans Scattering Theory

The simple Rayleigh scattering theory is based on the assumption that the relative refractive index of the scattering particle is close to unity and that the particle is much smaller than the wavelength of light. This may be briefly stated as

$$|m - 1| \ll 1 \tag{33a}$$

$$\frac{4\pi L}{\lambda} \ll 1 \tag{33b}$$

where characteristic linear dimension of the scattering particle is indicated by L; λ is the wavelength of light in the medium; and m/n_0 is the relative refractive index of the particle to that of the medium n/n_0. In this limit, the scattering of vertically polarized light is independent of angle.

The Rayleigh–Gans theory applies to larger particles but is also based on the concept of the oscillating electric dipole as is the simple Rayleigh theory. In this treatment, eqs. (33a) and (33b) are replaced by

$$\frac{4\pi L}{\lambda}|m - 1| \ll 1 \tag{33c}$$

In our experiments, in which large particles ($4\pi L/\lambda > 50$) are suspended in arbitrary media, eq. (33c) can be satisfied only by selecting $|m - 1|$ to be sufficiently small. As a practical matter, this is not an easy condition to satisfy.

The theoretical scattering pattern for large particles when eq. (33c) is no longer satisfied is very much in the forward direction in accord with that obtained from Rayleigh–Gans theory, but the details of the pattern are considerably modified. Roughly, this arises from the phase lag of the ray passing through the particle with respect to the ray going through the surrounding medium. This has been the subject of famous classical studies by Mie[18] and Debye[19]. More recently, a masterful analysis of this case has been presented for large spheres by van der Hulst[6]. The application of this analysis to the particular experiment described herein has not been carried through, and the reader is cautioned that the reliability of the interpretation is very strongly dependent on the extent to which eq. (33c) is obeyed.

4. DESCRIPTION OF THE PHOTOMETER

4.1. Outline of Problem

If the angular distribution of scattered intensity is to yield information concerning the arrangement of matter in the scattering sample, the range of angle to be examined depends on the structure of the scattering material. A length L characteristic of the scattering sample determines the wavelength and angular region which will provide the desired information. In general, one works with a wavelength λ, which is of the same order of magnitude as the characteristic dimension L, and the ratio λ/L indicates the angular region in which useful data may be obtained. The widely used techniques of large-angle light scattering are useful, e.g., in determining the size and shape of macromolecules in solution. These methods may be applied only to particles whose sizes are within certain limits. If the particles are too small, the scattered intensity will not exhibit the necessary dissymmetry effect. If the particles have dimensions up to several hundred times the wavelength of light, the scattering is limited to angles in the neighborhood of the incident beam. The study, then, of large colloidal particles and the structure of systems whose characteristic length is in the micron range necessitates the application of small-angle light-scattering techniques.

Toward this end a new, small-angle light-scattering photometer was constructed. The construction and a large part of the design of the instrument was undertaken by Natal Rao[20] of the Rao Instrument Co. The instrument employs photoelectric detection and is capable of both large- and small-angle measurements. Meaningful measurements to within 5′ of the incident beam are made possible by an optical system utilizing a focusing technique involving parallel incident light.

4.2. Optical System

The optical system is based in large measure upon a design[21] previously employed by Stein *et al.* The system is shown in figure 2. A 200-W Osram high-pressure mercury arc lamp (1) is enclosed in an air cooled chamber. A pair of heat-resistant 25 mm focal length lenses (2) focuses the brightest portion of the arc onto a pinhole 0.15 mm in diameter (4). In the parallel portion of the beam, between these lenses, neutral density filters (3) are inserted to attenuate the beam. The divergent beam, emerging from the pinhole, is made parallel by a 176.7-mm focal length double-convex lens (6). The lens is a good quality Bausch and Lomb lens, free from noticeable cosmetic defects and treated with antireflection coating (Balcoated). The diameter of the lens is 40 mm; however, the light stops (5), shutter, and iris (7) stop down the lens so that only one-twelfth of the lens is utilized. This provides a highly parallel beam, with a divergence of less than 5', and a circular cross section of 1 cm^2 where it strikes the scattering sample (8).

Mounted on the receiver arm is an identical 176.7-mm focal length lens (6), which at zero degrees refocuses the unscattered light through the exit pinhole (9), 0.15 mm in diameter. Situated behind this final pinhole are Corning compound filters (10) which isolate either the blue (4358 Å) or green (5461 Å) lines and a selected IP21 photomultiplier tube (11). The receiver arm is similarly equipped with shutter and iris (7) and light stops (5). For polarization measurements, polaroids are inserted at the entrance and exit of the sample chamber (12).

4.3. Mechanical Structure

The photometer has a rigid cast aluminum base 40 by 11 by 3 in. All the optical components are mounted on an optical bench track that allows

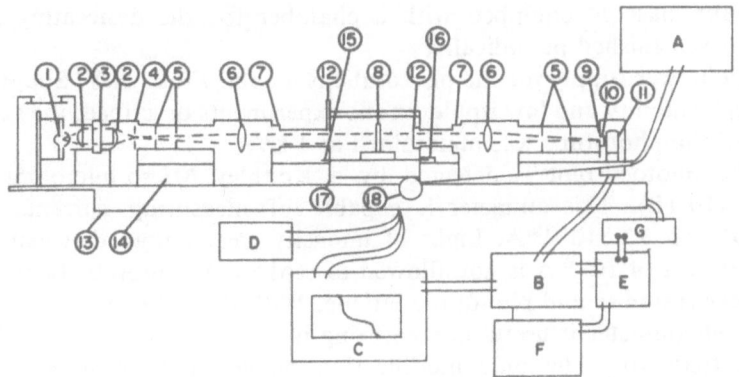

Figure 2. Layout of low-angle light-scattering photometer.

for a high degree of versatility in design and interchangeability of components. The detector is mounted on a sturdy cast aluminum arm (14), 19 in. long, which can rotate from -135 to $135°$. The beam, from entrance pinhole to photomultiplier, is enclosed in blackened tubing. Even as the arm is rotated the beam is enclosed by means of a flexible spring metal band encircling the sample chamber and attached to the detector arm. This avoids the necessity of maintaining the photometer in a light-tight box.

The angular position of the detector arm may be read directly from the dial (15) attached to the sample chamber. A vernier (16) allows the angle to be read to $1'$ of arc. The receiver arm may be moved by hand or by engaging a clutch (17), its movement is controlled by a knob (18) on the base of the photometer. By means of a worm-gear assembly, this knob controls small movements of the detector. One complete turn of the knob moves the detector through an angle of $1°$. The knob has a dial divided into 60 divisions, but because of a $10'$ backlash in the worm gear, this dial reads accurately only after take-up of the backlash.

The sample chamber has a diameter of 5 in. and an inside height of 4 in. The sample table will accommodate the usual solution cells or an adapter for thin filter samples. An extension in height of the chamber to 6 in. allows room for most tall cells used in solution studies. The cover and bottom of the chamber as well as its outside covering shell are made of bakelite, which facilitates future temperature control of the chamber.

4.4. Electronic Components

The photomultiplier tube is a specially selected RCA IP21 with a high signal-to-noise ratio. The phototube mount is mica-filled bakelite and wired with carbon-deposited resistors with 1% accuracy. When impressed with 1 kV the base leakage current is less than 10^{-13} A. The mount of the phototube base is equipped with a chamber for the desiccating agent which is replenished periodically.

The power supply for the phototube is a John Fluke H.V. Supply (A) with high stability and low ripple. In the experiments described, the voltage supplied the phototube was maintained at 1 kV.

The photocurrent is detected by a Keithley Micro-microammeter, model 410 (B). This ammeter is capable of measuring currents from 1×10^{-3} to 3×10^{-13} A. Light of intensity greater than that causing a photocurrent of 10^{-6} A is not allowed to strike the phototube because of nonlinear response and phototube fatigue. With the aid of neutral density filters, the ammeter is useful in measuring intensities that vary by a factor greater than 10^9. The only modification made on the Keithley Micro-microammeter is the addition of a capacitance across the input. A 15,000 pF capacitance shunted across the high-input terminal and ground necessarily reduces the response speed, but it also decreases the response to spurious a–c signals that interfere with the measurements. This additional capacitance was found to be the most satisfactory compromise between diminished noise and increased response time.

It was found that after a warm-up period of about half an hour the dark current becomes essentially constant and can be subtracted electrically by means of a simple bucking circuit of dry cells and a series 10^{12}-ohm bucking resistor.

4.5. Recording and Programming

The micro-microammeter is provided with output terminals. The output for full-scale meter deflection is 5 V and the maximum current that may be drawn from the output terminals is 5 mA. This output is sufficient for driving the pen channel of a Mosely X–Y Autograph recorder (C).

The shaft of the knob which controls the detector arm travel extends out the rear of the photometer base. By means of a belt drive and gear mechanism the detector arm may be driven at angular speeds ranging from 0.03 to 1°/min by a $\frac{1}{4}$-horsepower motor, mounted on the rear of the photometer base. The gear system also drives three 1K helical potentiometers; a one-turn continuous turning Heliport that rotates through 360° for every degree that the detector arm sweeps out, a ten-turn potentiometer that also rotates 360° for every degree that the detector arm moves, and a second ten-turn potentiometer that rotates 36° for every one receiver arm degree. These potentiometers are powered by a stable, low-ripple 10-V d–c supply (D). The output of any potentiometer may be used to drive the drum channel of the Mosely recorder. In this way the choice of potentiometer determines whether full-scale drum travel corresponds to a 1, 10, or 100° sweep of detector arm. The Mosely recorder thereby affords a direct record of intensity as a function of angle.

Since the micro-microammeter has 20 overlapping ranges from 1×10^{-3} to 3×10^{-13} A, in order to cover a significant angular region in a single record it is necessary that the rotary switch which selects the current range be turned automatically. To accomplish this the output of the ammeter is fed to an auxiliary 5-V full-scale deflection meter (E). This meter is equipped with relays which are fired when the meter pointer falls below or above two pre-set values. These meter relays are used to fire two rotary solenoids powered by a 110-V d–c supply. These solenoids (F) turn the shaft coupled to the rotary switch of the Keithley micro-microammeter. In this way, as the detector arm moves to larger angles, the intensity falls; when it falls sufficiently to cause contact at the low side of the meter relay, a solenoid is fired which switches the micro-microammeter to its next more sensitive scale. The other solenoid switches to the next lower scale when the intensity increases.

The operation of the photometer may be programmed by means of a double bank of slide switches (F) connecting the rotary switch and motor controller (G).* It is possible to pre-set these slide switches so that the detector arm will rotate at any one of three speeds or will stop, depending upon the scale selected by the rotary switch. In angular regions where the data are not important, or the intensity is slowly changing, high scanning rates are selected. In angular regions where the intensity varies rapidly, or on scales whose response times is long, a slow scanning rate is pre-set. In addition, the detector arm travel can be pre-set to stop at any scale, so that it will turn off if the light intensity becomes too high for the phototube or too low to be recorded. It is not necessary to know the actual rate of detector arm rotation, because the recorder plots intensity as a function only of angle and not of time.

4.6. Instrument Performance

Since detection is made without the use of a monitoring phototube, stability of the light source is an important prerequisite. The transformer which powers the Osram lamp is itself run from a Sola transformer which evens out the line fluctuations. Since the arc is brightest at one electrode, this region is focused at the pinhole, thereby diminishing arc wanderings. Steady measurements may be obtained if one waits an hour after the lamp is turned on. Although experimental reproducibility from day to day is obtainable only on a relative basis, over periods of hours the scattering envelope of an unmoved sample is reproducible within 1%.

With no scattering sample in the beam, the intensity falls by a factor of 10^4 as the receiver arm is moved from zero through an angle of 3'. But although the instrument is designed specifically for small-angle scattering measurements, it is also applicable to the more usual high-angle measurements because the receiver arm may rotate to an angle of 135°. Light scattered at large angles from pure liquids and dilute solutions is of such low intensity that the optical system must be modified. The initial pinhole is replaced by a circular hole 3 mm in diameter, and the final pinhole is replaced by an adjustable rectangular slit. By means of light stops the beam cross section is made square (1 by 1 cm) as it strikes the sample. Clearly, since the point source has been replaced, the beam's divergence is con-

---

* The motor and controller are manufactured by Gerald K. Heller Co., Las Vegas, Nevada. The knob turning mechanism was constructed by Daco Instrument Co., Brooklyn, New York.

siderably increased; but for high-angle measurements beam collimation is not as demanding as it is for small-angle measurements.

The light scattered from pure benzene (double distilled over sodium and filtered) and pure carbon tetrachloride (triply distilled over P_2O_5 and filtered) was measured at 90° to the incident beam (for which exit from the sample chamber was provided). The Rayleigh ratio of benzene obtained by Carr and Zimm[22] was used as the standard. The Rayleigh ratio of the carbon tetrachloride was then measured at two wavelengths. Despite the fact that the light intensity was very low, a close correspondence with previous results was obtained. The value for CCl_4 obtained at 4358 Å was 14.9×10^{-6} as compared with the 15.8×10^{-6} of Carr and Zimm; a value of 6.12×10^{-6} at 5461 Å agrees with the Carr and Zimm result, 5.88×10^{-6}. The internal consistency at these two wavelengths is good. The scattering of light is inversely proportional to the fourth power of the wavelength. Therefore, 6.12×10^{-6} multiplied by $(5461/4358)^4$ should equal 14.9×10^{-6}. The actual value is 15.1×10^{-6}.

The intensity of light scattered from polystyrene solutions of different concentration was measured. From the usual experimental plot the intercept was used to determine molecular weight. The value obtained was found to agree well with the values obtained on other photometers.

Samples used for low-angle measurement were prepared in the form of thin films. The discrete particles were dispersed in the desired matrix and sandwiched between clean microscope slides. Cover glasses, 0.015 to 0.020 cm thick, were used as spacers between the slides, and the edges were sealed. In all cases blanks were prepared from the pure matrix material. Considerable care was necessary to avoid trapping air bubbles in the samples and blanks. The intensity of scattered light at any angle was taken to be the intensity of light scattered by the sample in excess of the scattering from the blank.

Corrections were made for refraction effects. When light is scattered to an angle θ in a medium of refractive index $n > 1$, upon emerging from the medium the light is bent further away from the normal to the surface. The light appears to be scattered to an angle δ which is related to the true scattering angle θ by the simple expression $\sin \delta = n \sin \theta$.

Photometers which employ a convergent beam[23] have been shown to exhibit diffraction effects when the beam's dimensions are small. These effects can interfere with scattering measurements. In this photometer diffraction effects do not interfere with measurements. Possible diffraction rings caused by the small initial pinhole are eliminated by light stops along the long path lengths employed, and possible diffraction effects due to the beam where it strikes the sample would occur at immeasurably small angles because of the large beam cross section.

5. EXPERIMENTAL STUDIES

5.1. Materials

The scattering experiments were carried out on suspensions of glass spheres and of polystyrene latices. Two types of glass spheres were used. One type was supplied by the Minnesota Mining and Manufacturing Company. These were fractionated by air elutriation and sedimentation in water. The fractions were designated by LF and LS, respectively. A glass sample of smaller particle size was prepared in this laboratory, using a method similar to that of Bloomquist and Clark[24] with fine pyrex powder as a raw material. This sample was fractionated by sedimentation in water and designated by SS. The refractive indexes of the glass spheres were determined microscopically using the Becke line technique[25]. It was 1.515 for the LF and LS samples and 1.44 for the SS sample.

Seven samples of polystyrene latices of average particle size varying from 0.088 to 3.0 μ were obtained from the Dow Chemical Co.* Electromicrographs of the two large latices F and G are shown in figures 3 and 4 and appear to be of a high degree of uniformity. The refractive index of the latices was determined to be 1.59.

5.2. Preparation of the Light-Scattering Sample

The glass spheres were suspended in "Buton"—a styrene–butadiene copolymer with a refractive index of 1.534 in concentrations of 1 to 4% by weight. This was sandwiched between glass slides using cover glasses of 0.15 to 0.20 mm thick as spacers. Blanks containing the matrix alone were prepared, using the identical method. Care was taken to avoid trapping air bubbles in the samples and blanks.

The latices were suspended in glycerine or other matrices, which are mentioned later and were placed in a reentrant quartz spectrophotometer cell with a path length of 0.1 mm for measurement.

5.3. The Light-Scattering Data

The intensity of light scattered was recorded in arbitrary units as a function of angle for both the scattering samples and blanks of the pure matrix. The excess scattering of the sample over that of the blank was determined over an angular range from a few minutes to about 30° or until the intensity fell below the measurable limits. Most of the data were examined by plotting the intensity $I(h)$ vs. h, where $h = 4\pi/\lambda \sin(\theta/2)$ and λ is the wavelength of the light in the matrix, i.e., $\lambda = \lambda_0/n$. The measurements were carried out using the mercury blue line, $\lambda_0 = 4358$ Å.

---

* The authors owe thanks to Dr. John Vanderhoff for supplying the latex samples.

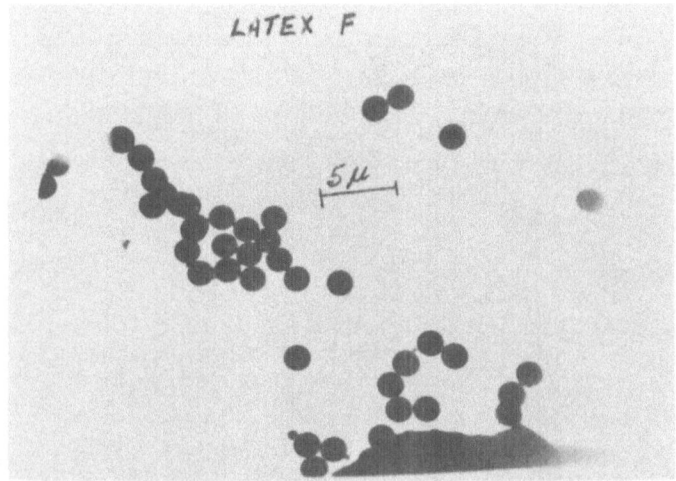

Figure 3. Electron micrograph of polystyrene latex F.

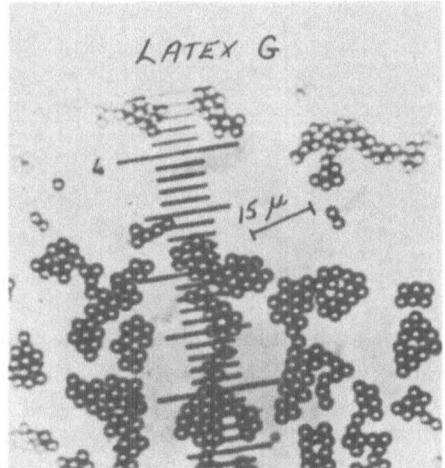

Figure 4. Electron micrograph of polystyrene latex G.

5.4. Experimental Results

5.4.1. *Glass Spheres*

The light-scattering results on the glass spheres were plotted according to Guinier's equation, eq. (15a) [log $I(h)$ vs. θ^2], and Debye's equation, eq. (23) [$I(h)^{-1/2}$ vs. θ^2]. Typical plots are shown in figures 5 and 6. The slopes of the Guinier plots yield the radii of gyration of the dispersed

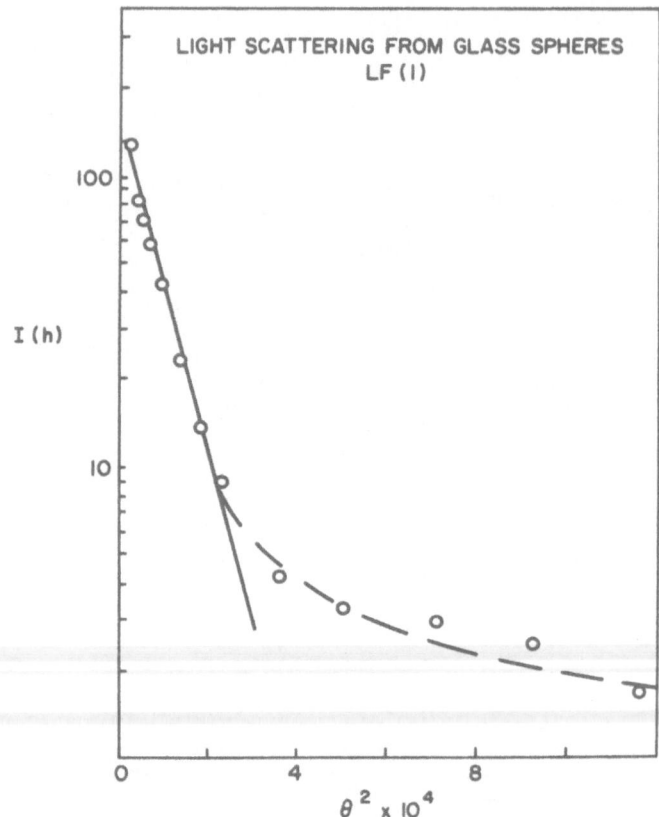

Figure 5. Guinier plot of light scattering from one glass sample.

particles. At high angles the Guinier plots deviate from linearity and become concave upward, characteristic of a polydispersed system. The particle sizes derived from the initial slopes are shown in table I. The average
particle sizes obtained from direct counting using photomicrographs and
electron micrographs are also included for comparison. The Debye plots
are not linear throughout the entire range measured, owing to the polydis-
persity of the sample in part. Where it is possible, several straight lines were
drawn, each covering a small angular region. The apparent surface areas
derived from these linear segments are also listed in table I. At least three
specimens were measured for each glass sample, and the results are reproducible.

The light-scattering data for the glass spheres were also treated by
the method of Hosemann[7] and Shull and Roess[8]. As discussed earlier,
these two treatments are based on an assumed distribution of particle size.

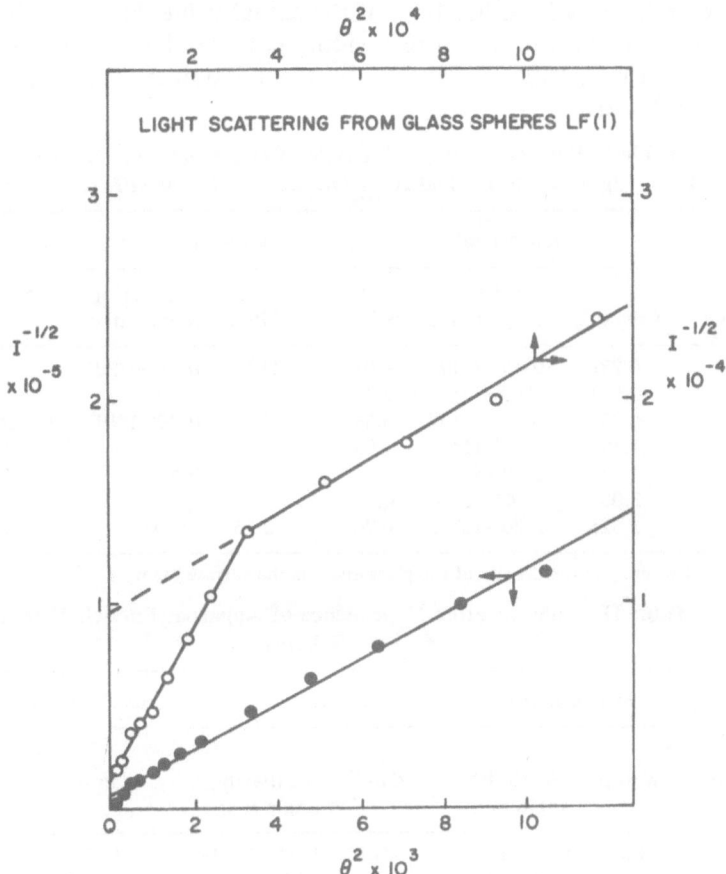

Figure 6. Debye plot of light scattering from one glass sample. ●, ordinate on left, abscissa on bottom; ○ ordinate on right, abscissa on top.

For details of these methods, the reader is referred to the original articles of Guinier and Fournet[4].

The effect of concentrations of glass in the matrix on the light-scattering results can be illustrated by the results from two samples, one containing 2% of glass and the other 30%. The apparent particle size as determined by the Guinier, Debye, and Hosemann methods show no strong concentration dependence. The results are shown in table II.

The surface areas of the glass spheres were also measured by Kr adsorption at $-195°$ and using the BET equation. This method has been widely adopted for measuring surface areas of fine particles, and the generally accepted reliability is within 10% of the actual value. Since the density of the particles is known, the corresponding particle diameter may be calculated from the specific surface $S_{\rm sp}$ by the relation $d = 6/(\rho S_{\rm sp})$, and are listed in table III.

Table I. Surface Area and Particle of Glass Spheres Obtained from Light Scattering and from Optical and Electron Microscopy

		Debye method			Guinier method		
Sample	Area m²/g	Scattering angle range	d^a in μ	d in μ	Scattering angle range	Microscope d in μ
LF(1)	0.288	0°15′–1°40′	8.0	23.5	0°15′–1°20′	18 ± 2
	1.071	1°40′–15°	2.15			
LS(6)	0.335	0°15′–1°40′	6.89	24.5	0°20′–1°10′	16 ± 5
	1.195	1°40′–15°	1.93			
SS(2)	1.218	1°10′–4°	1.89	3.26	0°50′–2°	0.5–5
	3.08	4°–10°	0.75			
SS(6)	3.525	1°20′–15°	0.76	2.15	3°–10°	0.2–2

^{a} $d = 6/(\rho s)$ where ρ is the density of the glass and s is the surface area.

Table II. Concentration Dependence of Apparent Particle Size of Glass Spheres

	Debye method		Guinier method		Hosemann	Microscope
Concentration (vol. %)	d in μ	Scattering angle range	d in μ	Scattering angle range	d in μ	d in μ
2	8.0	0°12′–1°40′	23.5	0°15′–1°20′	16.2	20.1
	2.15	1°40′–15°				
30	11.9	0°20′–2°	19.8	0°20′–1°20′	18.6	20.1
	3.6	1°40′–7°				
	2.0	7°–15°				

Table III. Surface Area of Glass Spheres by Kr Adsorption (BET Method)

Sample	Microscope d in μ	BET area (m²/g)	BET d in μ
LF	20.1	0.178	13.0
SS	1.6	8.12	0.35
SS	1.0	11.95	0.20
SS	2.0	2.94	0.80
LS	22.7	0.121	19.1

In comparing the microscopic and BET d values one should take into consideration the following factors:

1. The microscopic d is the arithmetic mean of the diameter of the particles as seen under the microscope

$$\bar{d} = \frac{1}{N} \sum_{i=1}^{N} d_i$$

while that derived from the BET equation and gas adsorption is statistically weighted according to

$$d = \frac{6W}{\sum_{i} n_i d_i^2}$$

2. Microscopic counting generally results in too large a value for d since some small particles are missed.

3. Glass spheres, especially after being exposed to water, have very rough surfaces.

The gas-adsorption result is a fair approximation for the true surface area, and the actual particle size is only slightly less than that derived from microscopic counting.

Considering the nonuniformity of the particle size and the arbitrary distribution function used, the agreement between the particle sizes determined by the Guinier and Hosemann methods with that from microscopic counting is not unreasonable.

The Debye d values in general are much less than the others. This is not unexpected as the surfaces of the glass spheres are rough. Furthermore, the nonlinearity of the Debye plots reflects the wide distribution of particle size, the d value (or surface area) derived is dependent on the angular range used. At high enough angle, the Debye surface area is expected to be the same as that determined by gas adsorption. But from the present data no clear-cut criteria can be adopted to make the comparison possible.

5.4.2. Polystyrene Latices

The Guinier plots, log $I(h)$ vs. h^2, for the seven latices are shown in figures 7 through 11. The plots are not linear for the four small latices owing to their wide distribution in particle size owing to agglomeration, while those of the three large latices are linear over the region for which the Guinier approximation is valid. The particle sizes derived from the initial slopes are listed in table IV. It is shown that only the three large latices E, F, and G give radii comparable to those obtained from electron micrographs. The agglomeration of particles in latices A, B, C, and D

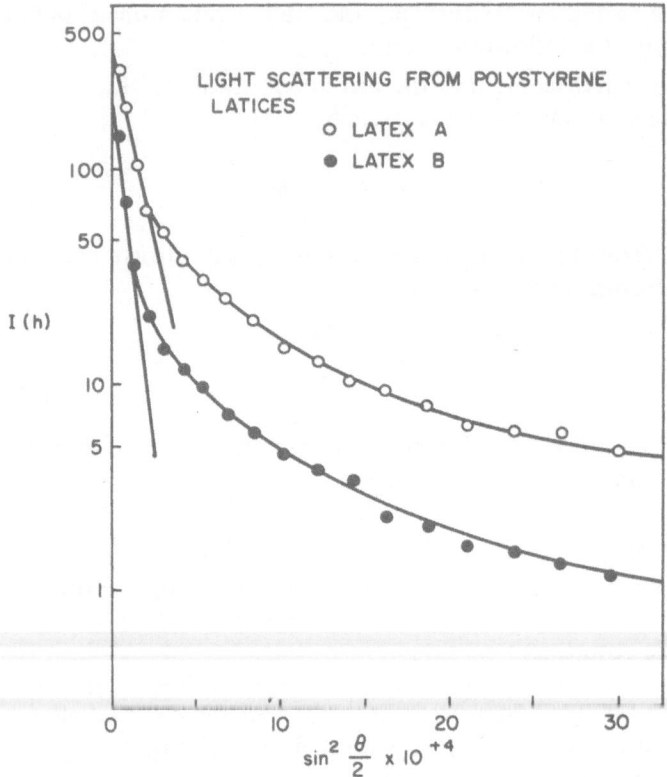

Figure 7. Light scattering of latices A and B as a function of angle.

Table IV. Particle Sizes of Polystyrene Latices[a]

Latex	Microscope d in μ	Guinier d in μ
A	0.088	4.85
B	0.264	4.07
C	0.557	3.34
D	0.814	5.88
E	1.17	1.31
F	1.80	1.77
G	3.0	4.23

[a] Measurements in suspension in glycerine. Much clustering of latices A through D observed in electron-microscope photograph.

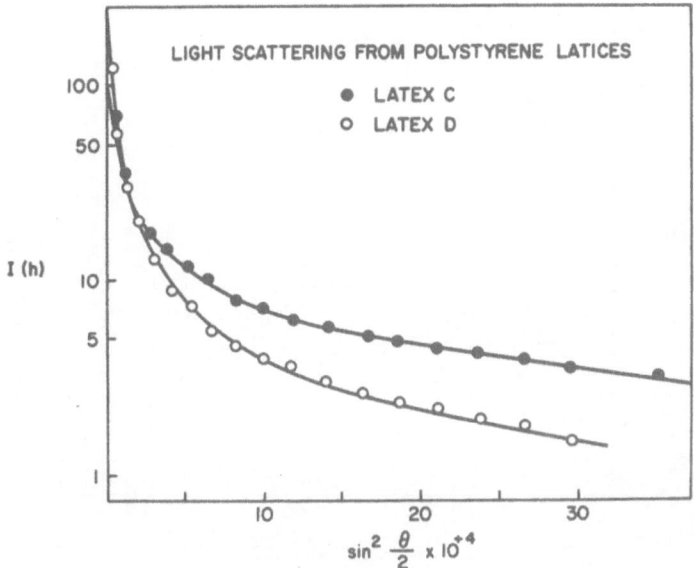

Figure 8. Light scattering of latices C and D as a function of angle.

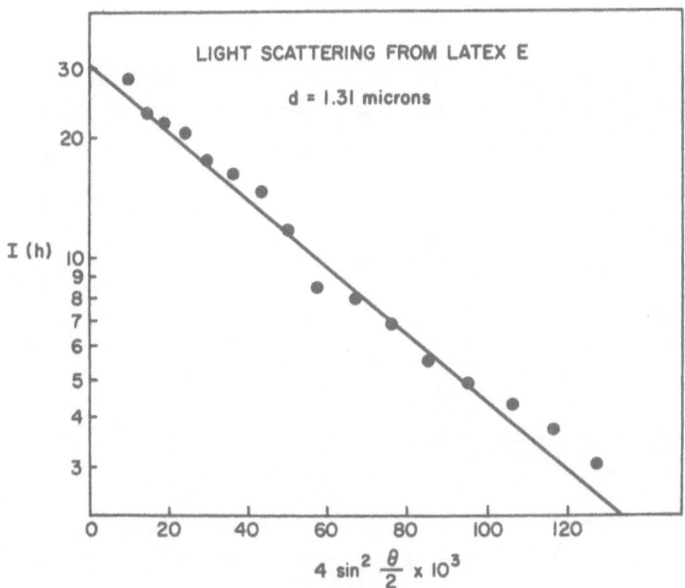

Figure 9. Light scattering of latex E as a function of angle.

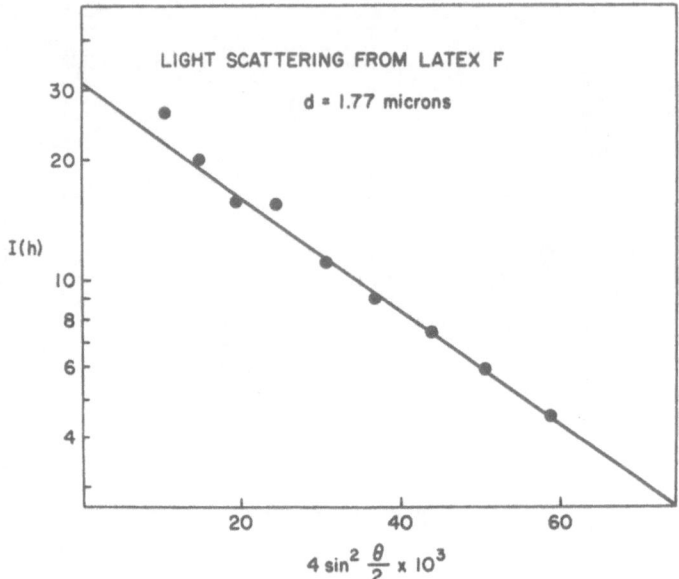

Figure 10. Light scattering of latex F as a function of angle.

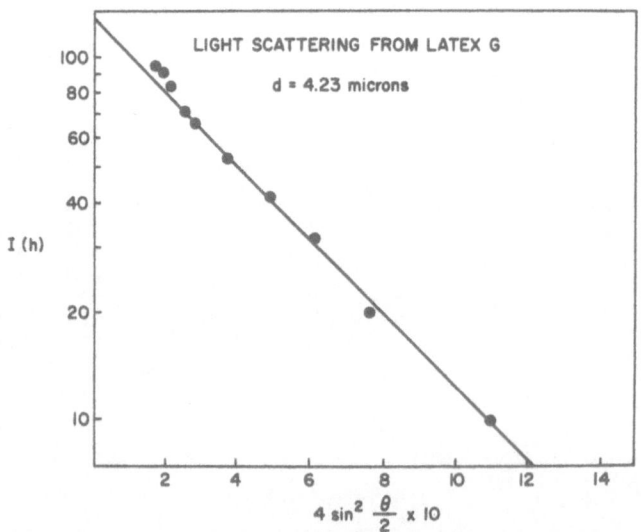

Figure 11. Light scattering of latex G as a function of angle.

gives an initial slope of Guinier plot much too high. Therefore, latices F and G, which are reasonably well dispersed, are chosen for further studies.

The latices F and G are suspended in the following matrices:

1. Water–gelatin solution, $n = 1.342$
2. Glycerine–gelatin solution, $n = 1.484$
3. Glycerine–phenol solution, $n = 1.506$

The samples show maxima and minima in their scattering envelopes. The intensity of light scattered from a dilute suspension of spheres is given by

$$I(h) \sim \left[\frac{\sin hR - hR \cos hR}{(hR)^3}\right]^2$$

On this basis the location of the intensity maxima is determined by

$$\tan hR = \frac{3hR}{3 - h^2R^2} \tag{35}$$

The plot in figure 12 shows that the maxima occur at the values of $hR = 5.76$ 9.10, 15.52, 18.69, or approximately $hR \simeq n\pi$, where $n = 2, 3, 4, 5, \ldots$. For small hR, $3hR/(3 - h^2R^2) < \tan hR$; therefore, no maximum exists at $hR \simeq \pi$. The apparent radii calculated from the location of the peaks for latex G in the three matrices are listed in table V.

The Guinier and Debye plots of latices F and G in the three matrices led to values of the radii listed in table VI. Diameters measured in the dry state from electron micrographs are included. Considering that the microscopic values were derived from dry latices, and some swelling of the

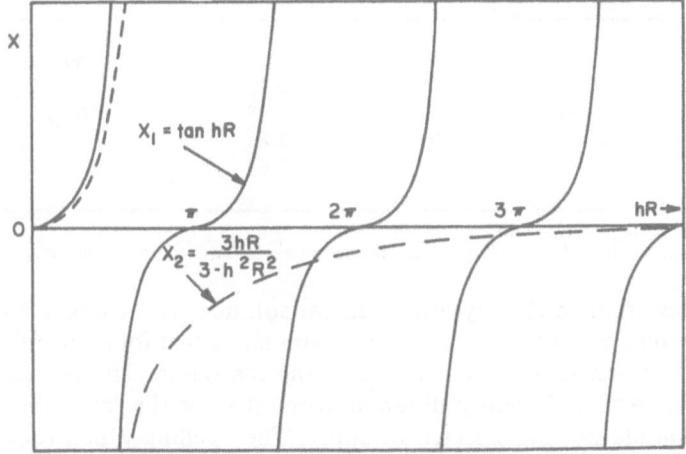

Figure 12. A graph showing the location of the roots of the equation $\tan hR = 3hR/(3 — h^2R^2)$.

Table V. Apparent Radius of Latex G from Positions of Maxima in Scattering Curve[a]

		Position of maximum	
Order of maximum	Matrix	$2 \sin \theta/2$	Radius in μ
1	A	0.1532	1.94
1	B	0.1326	2.04
2	A	0.3006	1.57
2	B	0.2395	1.85
3	A	0.4116	1.54
3	B	0.3295	1.76
3	C	0.3195	1.78
4	A	0.5193	1.54
4	B	0.4189	1.73
4	C	0.4015	1.76
5	A	0.6136	1.56
5	B	0.5076	1.72
5	C	0.4893	1.76

[a] Samples measured in matrices of varying refractive index, $n_A = 1.342$, $n_B = 1.484$, and $n_C = 1.506$.

Table VI. Particle Size of Some Polystyrene Latices[a]

Latex	Matrix	Debye method d in μ	Guinier method d in μ	Microscope d in μ (dry Latex)
F	A	0.84	2.44	1.80
F	B		2.12	
G	A	1.60	3.68	3.0
G	B	1.60	2.88	
G	C[b]		3.64	

[a] $n_A = 1.342$, $n_B = 1.484$, and $n_C = 1.506$.
[b] Some spheres in matrix C, a glycerine phenol mixture, swelled to occupy larger diameters.

latices occurred in the glycerine–phenol solution, the Guinier and microscopic values are in reasonable accord, similar to that found for glass beads. This conclusion is based on the following reasoning. The refractive index of matrix A is sufficiently different from that of the latex that the RG theory should not be expected to apply. The swelling which took place in matrix C would lead to radii determined by the Guinier method which were larger than the results obtained in the microscope. The results in matrix B were in reasonable agreement with that obtained from the electron micrographs. The small Debye d values, corresponding to large surface area, are expected as the surfaces of the latices are by no means smooth.

The light scattering from a mixture of latices F and G (weight ratio 1 : 1) in water gelatin ($n = 1.34$) was also determined. As is shown in figure 13, the Debye plot of this system consists of two segments. The low-angle region parallel to that of the larger latices (G) and the high-angle region is predominantly determined by the smaller latices (F).

6. DISCUSSION

The light scattering of suspensions of spheres in various matrices was measured using a low-angle light-scattering photometer. The results of the measurements were reproducible; the analysis of the measurements by varying techniques demonstrated that the apparent result is very sensitive to ad hoc assumptions about the form of the correlation function or the nature of particle-size distribution.

The fact that the interpretation of the results was based on the Rayleigh–Gans theory limits the reliability of the analysis. This is particularly true whenever the particle size becomes large and the refractive ratio of particle to matrix deviates from unity.

Two important developments since these experiments were completed (1962) would make the problem much simpler than it was. The first of these

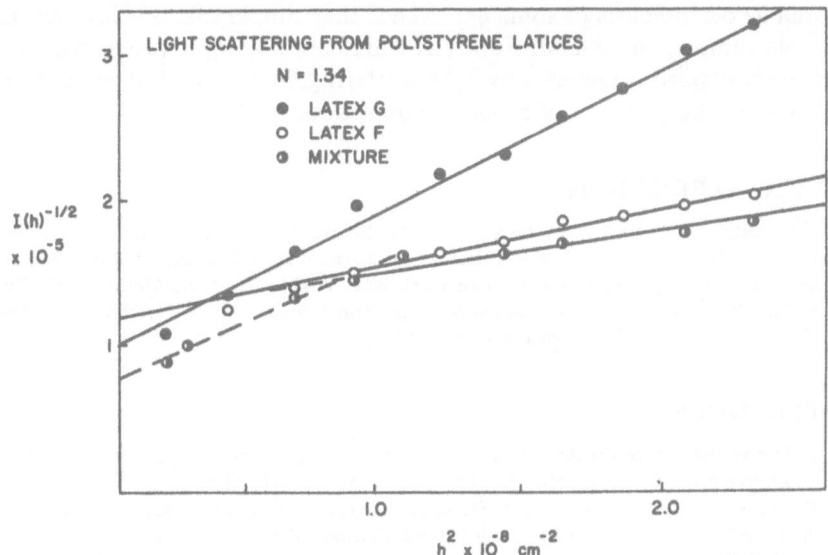

Figure 13. Debye plots for latices F and G and a 1 : 1 mixture of the two.

is the widespread availability of high-speed computation facilities which would make it possible to analyze the results of an experiment using calculated correlation functions and their integrals and derivatives. The second is the development of the continuous gas laser which emits a parallel beam permitting measurements at a small angle with a reasonably intense source of light.

The key problem which the research was originally initiated to solve was the utilization of light scattering for the analysis of porous minerals. Some attempts on actual samples not discussed in this paper but contained in a doctoral dissertation of D. Caulfield on which this work is largely based[23] were not easy to handle because of the absorption of light. We believe that if transparent replicas of porous minerals could be made so that the two component mixtures accurately reproduced the geometry of the sample, scattering methods would be feasible for structure determination provided that the refractive indices of the two components were not too different from each other.

The studies made here and elsewhere have persuaded us that light-scattering measurements can be carried out at low angle which are capable of systematic analysis, provided that the scattering can be described properly by RG theory.

The problem which remains is the preparation of samples or replicas. We have no information on whether suitable replicas could be prepared of samples of interest, though we are reasonably sure on theoretical grounds as well as on the basis of some experience that simple rock sections are not suitable samples in general. Future researchers, hoping to study the structure of these porous minerals by light scattering, in our opinion, should concentrate on the problem of replica preparation.

ACKNOWLEDGMENTS

The authors received financial support for this work from the American Petroleum Institute and the Petroleum Research Fund of the American Chemical Society. Technical advice and strong moral support for this work was provided by T. M. Geffen of the Pan-American Petroleum Co., F. L. Johnson of the Sun Oil Co., W. R. Purcell of the Shell Oil Co., and M. R. J. Wyllie of the Gulf Oil Co.

REFERENCES

1. P. Debye, Report to the American Petroleum Institute, New York, N.Y. 10020 (1955).
2. P. Debye and A. M. Bueche, *J. Appl. Phys.*, **20**, 518–525 (1949).
3. P. Debye, H. R. Anderson, and H. Brumberger, *J. Appl. Phys.* **28**, 679–683 (1957).
4. A. Guinier and G. Fournet, *Small-Angle Scattering of X-Rays*, John Wiley, New York (1955).
5. G. Porod, *Kolloid-Z.* **124**, 83 (1951); **125**, 109 (1952).

6. H. C. van der Hulst, *Light Scattering of Small Particles*, John Wiley, New York (1957).
7. R. Hosemann, *Ergeb. Exakt. Naturw.* **24**, 142–221 (1951).
8. C. G. Shull and L. C. Roess, *J. Appl. Phys.* **18**, 295–307 (1947).
9. M. H. Jellinek, E. Solomon, and I. Fankuchen, *Ind. Eng. Chem. Anal. Ed.* **18**, 172–175 (1946).
10. S. H. Bauer, *J. Chem. Phys.* **13**, 450–451 (1945).
11. L. C. Roess, *J. Chem. Phys.* **14**, 695–697 (1946).
12. J. Riseman, *Acta Cryst.* **5**, 193–196 (1952).
13. V. Luzzati, *Acta Cryst.* **10**, 33–34 (1957).
14. V. Luzzati, *Acta Cryst.* **10**, 643–648 (1957).
15. J. L. Soulé, *Suppl. J. Phys. Rad.* **18**, 90A (1957).
16. R. Baro and V. Luzzati, *Acta Cryst.* **12**, 144–148 (1959).
17. D. Caulfield and R. Ullman, *J. Appl. Phys.* **33**, 1737–1740 (1962).
18. G. Mie, *Ann. Physik*, 377 (1908).
19. P. Debye, *Ann. Physik* **30**, 59 (1909).
20. Rao Instrument Co., Brooklyn, New York.
21. R. S. Stein, A. Plaza, and F. Norris, *J. Polymer Sci.* **24**, 455–460 (1957).
22. C. I. Carr, Jr., and B. H. Zimm, *J. Chem. Phys.* **18**, 1616–1626 (1950).
23. W. H. Aughey and F. J. Baum, *J. Opt. Soc. Am.* **44**, 833–837 (1954).
24. S. Bloomquist and P. Clarke, *Ind. Eng. Chem. Anal. Ed.* **12**, 61–62 (1942).
25. E. E. Wahlstrom, *Optical Crystallography*, John Wiley, New York (1951).
26. Daniel Caulfield, PhD dissertation, Polytechnic Institute of Brooklyn, Brooklyn, New York (1962).

INDEX